CONTEMPORARY MOTHERHOOD

To Don, Jack and Cassie, with love

Contemporary Motherhood
The Impact of Children on Adult Time

LYN CRAIG
University of New South Wales, Australia

ASHGATE

Published by
Ashgate Publishing Limited
Gower House
Croft Road
Aldershot
Hampshire GU11 3HR
England

Ashgate Publishing Company
Suite 420
101 Cherry Street
Burlington, VT 05401-4405
USA

Ashgate website: http://www.ashgate.com

British Library Cataloguing in Publication Data
Craig, Lyn
  Contemporary motherhood : the impact of children on adult
  time
  1. Motherhood 2. Women - Social conditions 3. Family - Time
  management 4. Child rearing
  I. Title
  306.8'743

Library of Congress Cataloging-in-Publication Data
Craig, Lyn, 1955-
  Contemporary motherhood : the impact of children on adult time / by Lyn Craig.
       p. cm.
  Includes index.
  ISBN: 978-0-7546-4998-4 1. Child rearing--Study and teaching. 2.
Parenting--Economic aspects. 3. Time measurements. 4. Family--Research. I. Title.
  HQ755.7.C73 2007
  306.87409182'1--dc22

                                                                          2007002934

ISBN: 978-0-7546-4998-4

Printed and bound in Great Britain by Antony Rowe Ltd, Chippenham, Wiltshire.

Contents

List of Figures

List of Tables

Acknowledgements

I gratefully acknowledge the Australian Research Council and the Office for Women, Department of Family and Community Services and Indigenous Affairs, Australian Federal Government for the support provided through Discovery Grant/Post Doctoral Fellowship DP0665337, and the Time Use Fellowship 2004–5. This book presents results drawn from the Multinational Time Use Study (MTUS) and from the Australian Bureau of Statistics (ABS) Time Use Survey, and I take full responsibility for the interpretation of data and for any errors or omissions. The views expressed are my own, and do not necessarily represent those of the Australian Research Council, the Office for Women, the MTUS team, the ABS or the Australian Government. I also wish to express my gratitude to all those who have given me practical assistance, advice or encouragement, particularly Janeen Baxter, Michael Bittman, Jude Brown, Paula England, Nancy Folbre, Diana Harris, Kristy Muir, Deborah Oxley, Melissa Roughley, Peter Saunders, Denise Thompson and Don Wright. Thank you all.

Chapter 1

Introduction

Becoming a parent is not only one of the most significant rites of passage in the human life course but a contribution to the perpetuation of the human race. For such a fundamental event, parenthood has become remarkably problematic. On a social level, this is evident in delayed childbearing and falling fertility rates. Becoming a parent poses, for women at least, a considerable dilemma. One of the most perplexing contemporary challenges for young women is whether to become a mother, when to become a mother, and how to manage being a mother to children once born. How could something so basic to human society cause such a quandary?

One answer is that it is because we have had half a sex revolution. Gender roles are much more similar, in that women have entered the work force in huge numbers, and male and female contributions to domestic labour, though not equal, have become more so. However, these changes have occurred through a revolution in women's behaviour, not men's. Women have taken up paid work; women have reduced the time they devote to housework. In the face of this, men's behaviour has, relatively speaking, remained remarkably constant. The problem is that unilateral change in female behaviour is not adequate to the issue of children. There is a sticking point in the revolution: someone has to take care of the kids.

The problem is compounded by the fact that caring for children is largely invisible and no framework for social and economic accounting for it exists. A major cost of parenting is the time it takes. No adequate social provision for this time is being made; it is not quantified, and its extent is unknown. The issue is left to women to deal with individually, on an ad hoc basis, personally weighing the relative benefits and disincentives of remaining childless, embracing full time motherhood or 'balancing' work and family demands.

The lack of social recognition of the time demands of parenthood may mean that prospective mothers themselves are unaware of how large they will be. When new mothers meet the reality of the stalled revolution, there can be (along with exhaustion) feelings of confusion, bewilderment and betrayal. What they are doing is providing an essential and publicly useful service that is insufficiently supported by social institutions. This can be hard to pinpoint, not least because of the attachment women feel for their children. Lack of support does not mean they can withdraw the service, because their children need it. Just because it is not fair does not mean they will not do it. The opposite is also true: just because they do it, does not mean it is fair.

Also making the problem hard to articulate is that there is not a common social understanding of the issue. Attitudes to children, to gender issues and to paid work can operate on different planes. Nor does current social science theory clarify the matter. Children and parenthood are of interest to several disciplines, but none deals with all aspects of it. Psychology has promulgated the idea that sustained and attentive nurturing is essential to optimal child development. It puts great emphasis on the parent-child relationship, in which the children's needs are paramount, but it is silent on the social conditions necessary for this to occur. Sociology deals with the family once constituted, but not with factors that may influence whether people become parents and how many children they may have. Economics does address this aspect of parenthood, suggesting that as costs of children go up, fertility will go down. But economic theories offer little room for understanding how the costs of children vary by sex, of the effect of unequal power within families, or of the complexity of how a mother's 'utility' or welfare can be both in conflict with and interwoven with her child's. Feminist writers recognise that power is unequal by gender, and that women and children's interests can be both entangled and opposed, but a formulation whereby the needs of both could be fully met has yet to be articulated.

These partial and contradictory perspectives do not clarify the contemporary predicament for parents and prospective parents. None fully addresses the issue of how having children affects the lives of parents themselves. None asks, let alone answers, the question: what are the time effects of having children? In this book I investigate this hidden cost of parenthood through a comprehensive and detailed analysis of the impact of children upon adult time.

Background

One reason the time impact of children upon their parents has not been widely researched is that it is viewed from so many different perspectives that it is difficult to define the issue. Depending on the viewpoint, children are variously a private responsibility or a joint social concern, and women and children's interests are either commensurate, opposed, or parenting is an androgynous affair to which gender concerns are irrelevant. Of central importance is the question of who is regarded as being responsible for and benefiting from children. Economic and social policy perspectives on children range across extremes. At one end, as Nancy Folbre notes, neo-classical theorists see having children as a private indulgence, very similar to having pets (Folbre, 2001). At the other, it is seen as a social contribution to the economic well being of all (Chesnais, 1998). A brief historical overview can show some of the different and often incommensurate ways in which time with children has been framed.

Over the course of the Industrial Revolution, the productive family unit at the core of pre-industrial economic organisation was replaced by the (allegedly) independent wage labourer (Eisenstein, 1979). Male labour moved out of the household; women's labour remained in (or returned to) the household. The shift was reflected in political theory and in social attitudes. An idea that became prominent in western liberalism was that public and private constitute separate spheres, with men naturally suited to engage in the public sphere, and women appropriately being cloistered in the private (Pateman, 1988; England, 1993). The home came to be regarded as a private domain that stood outside the economy (Gilding, 1991; Fukuyama, 1999; O'Connor et al., 1999). This meant that a largely arbitrary line was drawn between paid and unpaid work. Household work became economically invisible. Economic theory defined work as only those activities undertaken for money, which meant unpaid household labour was categorised as leisure, rather than as productive activity (Folbre, 1991). Domestic labour was redefined. It was not work, but a dimension of femininity (Oakley, 1985). The household retained control over women's labour, but that labour was no longer recognised as productive (Cowan, 1983; Folbre, 1991).

Historically, children had also made an economic contribution to their households. Depending on social level, children brought to a family the possibilities of additional home labour, extra earnings, or land acquisition through dynastic connection (Aries, 1962; Shorter, 1977; Zelizer, 1985). Through the profound social changes of the last two centuries including but not limited to industrial working practices, child labour laws and compulsory education, the historically positive economic implications of having children were reversed (Schultz, 1974). Wealth no longer flowed within families from the young to the old and children became a net economic drain on family resources (Caldwell, 1982).

They also required time. Looking after children became a discrete, dedicated activity, which was conducted in the private domestic sphere. Over the course of the nineteenth century, social expectations of the parenting role changed considerably (de Mause, 1974; Shorter, 1977; Donzelot, 1979). The male breadwinner household, which was regarded as comprising an economically active man, who worked outside the home for money, and his economically inactive wife and dependent children, became not only standard but idealised (Casey, 1989; Gilding, 1991). The gender division of labour was widely viewed as reflecting natural, biologically based differences between the sexes (Oakley, 1985; Folbre, 1994b; Cass, 1995; Baxter, 2002). Women bore children, and were best suited to raise them. This essentialist view was strongly reinforced by cultural prescriptions about mothering, which have grown ever since the cult of domesticity and the sentimentalisation of childhood of Victorian times (Aries, 1962; Badinter, 1981; Dally, 1982; Casey, 1989; Ehrenreich and English, 1989).

The amount and type of care seen to be necessary for children expanded (Cowan, 1983; Reiger, 1985). Originally dominated by the belief that women should ensure children's physical well being (Donzelot, 1979), expectations of good mothering grew to encompass psychological considerations (Cowan, 1983). Theories of child development and psychology came to suggest that maternal bonding, attentive parenting and high time inputs are requisite for optimal educational and social outcomes (Bowlby, 1972; Bowlby, 1973; Belsky, 2001). It was thought paramount for individual development that the person delivering care to children was their own mother (Bowlby, 1953), which entrenched the similarity of women's responsibilities across social class (Cowan, 1983). The idea that good mothering required constant presence meant both that middle class women could no longer delegate and, conversely, that working class women were encouraged to stay at home to care for their children rather than to take employment.

The gender division of labour was largely unquestioned until the advent of second-wave feminism, when women began to claim a place in the public sphere (Oakley, 1985; Folbre, 1994b; Cass, 1995; Baxter, 2002). They grasped opportunities to acquire education, and developed aspirations to join the work force, to have careers and to be independent. Many changes ensued. Arguably, the most significant economic and social development since the Second World War has been the increase in the participation of women in the labour force (Mincer, 1962). Concomitantly, relations in the private sphere became open to question, and feminist critique identified the family, and responsibility for unpaid domestic labour, as a major source of female oppression (Firestone, 1970; Mitchell, 1971; Oakley, 1974; Delphy and Leonard, 1984).

However, motherhood was a particularly challenging area. On one view, the route to gender equity is blocked by motherhood and care responsibilities. Many feminist writers have suggested that 'maternal instinct' is an ideology that grew specifically to encourage women to accept the loss of independence commensurate with full time mothering (see for example Mitchell, 1971; Badinter, 1981). Mothering has been described as the point at which female oppression is most keenly felt (Nava, 1983). If having children makes women economically vulnerable, it provides a practical reason for a power differential between men and women. The glorification of motherhood and concomitant sentimentalisation of childhood was seen as a primary way in which women were coerced into accepting a subordinate position. Shulamith Firestone argued that the fact that women are potential child-bearers and men are not is the central cause of gender inequality, and that it could be overcome only by all women abdicating the reproductive role (Firestone, 1970). To achieve gender equality, women should not have children.

A problem with this solution to inequity is that many women do wish to be mothers. For many, not to have children would be a personal sadness. How are those that wish to, to have children without disadvantage? Answering this question is considerably more challenging than advocating a baby strike. Also, it is easier to dispute the assumption that all women are instinctively and innately capable of care than to repudiate ideas about what infants instinctively need from mothers (Hrdy, 1999). The idea that children need attentive nurturing is now thoroughly accepted, which means that many who promote female rights to independence are also concerned that if women substantially withdraw from care the welfare of children will suffer (Pfau-Effinger, 2000; Hewlett et al., 2002; Gornick and Meyers, 2004). Balancing the needs of children for nurture and the needs of women for independence without overloading women is an enduring challenge.

Rejecting motherhood altogether proved too radical for most, and the preferred route to equality became that of sharing the care. 'If we move from a gender-divided society to a more equal one, then we have to go the whole way if children are to be adequately cared for' (Himmelweit, 2000, p.2). Language became studiously gender neutral. The substitution of terms such as 'primary carer' and 'parenting' for 'mother' and 'mothering' opens the theoretical potential for others, including men, to join women in the care of children. However, this approach has potential pitfalls. If in fact the carers of children are mainly women, and/or if there are unequal power relations between the sexes, then making gender linguistically invisible is more obfuscatory than illuminating (Nava, 1983). A risk of a gender-free conceptualisation of parenting in a male dominated society is that it obscures the specific position of mothers.

This is to a large extent what has happened. The gender-neutral approach fitted the mood of the times. Most western countries have moved towards neo-liberalism, a restatement of classical liberalism, which asserts the liberal principles of freedom, market individualism and small government (O'Connor et al., 1999). Feminism and liberalism share a common origin, in that both are predicated in arguments of equality (Pateman, 1988; Coltrane, 1996; O'Connor et al., 1999). Once it was recognised that it was not theoretically consistent to exclude women from the precepts of liberalism, the principle of gender equity became increasingly incorporated into employment conditions and social policy, including family law (Cass, 1995; Shaver, 1995). Women now claim a place in the public sphere and are counted as individual agents. However, neo-liberalism also perpetuates the view that family arrangements should be private. Neo-liberalism defines both men and women as possessive individuals but sees the sexual division of labour in paid and unpaid work as a matter of private choice by marital partners (O'Connor, 1999). On this view, if people's choices in the workplace are constrained by their domestic responsibilities, that is a personal issue, not a social concern.

Neo-liberalism is intellectually underpinned by neo-classical economics, one of the core assumptions of which is that people will behave as self-interested individuals who seek to maximise their utility (satisfaction) through rationally choosing optimal outcomes. Utility is seen to be subjective, so interpersonal comparisons of welfare are not possible, and people are assumed to have constant, unchanging preferences. The choosing individual is not necessarily conscious of this choice process but their preferences are revealed by the outcomes of their behaviour (Samuelson and Nordhaus, 1985). This theory was originally used to explain or predict decisions related to the market, but theorists have extended its application to areas of social life once regarded as peripheral to the economy. Of relevance to this book, Gary Becker applied this type of economic analysis to the family in a formulation dubbed 'new home economics' (Becker, 1981; Becker, 1991). He 'brought the theoretical apparatus developed in the last hundred years for the analysis of markets and the explanation of economic growth to bear upon the household' (Bergmann, 1995, p. 142). New home economics retains the concepts of methodological individualism, rational choice and utility maximisation central to neo-classical economics by treating the household as a single unit, to which it attributes a joint utility function (Becker, 1981). The assumption is that an altruistic head of the household will make decisions in the best interests of all.

Building on the pioneering work of Margaret Reid (Reid, 1934), new home economics acknowledges domestic labour has economic value. Home production makes an economic contribution to family wellbeing, as does earning money in the market place. Becker's new contribution was to argue that dividing market and domestic labour by sex is economically rational; that it is efficient, yields the best return for effort and improves welfare by maximising gains to all (Becker, 1981; Becker, 1985; Becker, 1991). Men have a comparative advantage in market work, women have a comparative advantage in home production because they bear the children, and it is through specialisation that each partner best contributes to the welfare of the family unit as a whole (Becker, 1991).

The theory does not include caring for children within its definition of economic productivity. Although Becker acknowledges that child care can be demanding, he theorises that having children is like acquiring a consumption item (Becker,

1965; Leibenstein, 1974; Becker, 1985). Since children are an economic drain on families, he regards prospective parents as rationally choosing whether or not to have a child, by weighing up the costs and benefits as in other consumer choices. The major benefit of having children is the pleasure or utility obtained by the parents as a unit (Bergmann, 1995), and reflecting the essentialist assumptions of the theory that women will provide all household services, the major cost of children is the opportunity cost of the mother's time.

So having children is a decision akin to buying a new fridge or car, and as with other consumer items, demand will fall as costs rise. The theory predicts that people will have fewer children as they become more expensive, in particular that declining fertility will result when women make significant investments in their own human capital (Leibenstein, 1974). Households with educated women are likely to have fewer children because the opportunity costs of leaving the work force are higher for the more highly educated (Becker, 1981; Becker, 1991). Therefore, the theory argues, a household with an educated woman will have fewer children in order to direct more of her time to paid work. Resources will be concentrated on those fewer children. This, it is suggested, will mean an increase in the 'quality' of children concomitant with a reduction in the 'quantity' of children (Becker, 1981, p.110).

Parents are presumed to have rationally decided the costs of the children they do decide to have are outweighed by the emotional benefits. The presence of children, therefore, is a revealed preference for which compensation is unnecessary (Bradbury, 2001). Within the theory, the motivation, the costs and the rewards for having children are entirely private, and there is no need for government assistance or interference (Bergmann, 1995). How fathers and mothers allocate the care time involved is a private matter and, anyway, women are suited to it by nature. Combining these ideas means that the time cost of children and any difference in its impact by sex are defined out of social or economic enquiry. Having children is simply what some women want to do for fun. Any ramifications for labour supply or domestic equity are ignored. It is thought that partnerships are now more equal, can be repudiated if unsatisfactory (Giddens, 1991; Hakim, 2000; Beck and Beck-Gernsheim, 2002; Beck-Gernsheim, 2002), and women choose their allocation of time between work and paid work, based on what they prefer to do (Hakim, 2000). From this perspective, children are chosen, a private good, and any gap in time (or money) expenditures between parents and the childless, or between mothers and fathers, is not a matter of public concern.

An alternative economic view is that children are not a private, but a public good. Public goods are those that cannot be supplied to any one person without benefiting others, and the use of which by any one person does not preclude their use by others (Caporaso and Levinde, 1992). An example is lighthouses, which protect all ships from harm, including those that do not contribute to their cost (Bittman and Pixley, 1997). Unless contributions are gathered through a system such as tax collection, there are difficulties in getting beneficiaries to pay for public goods, thus allowing the emergence of 'free riders' who enjoy the benefit but do not pay for it (Caporaso and Levinde, 1992).

Children can be viewed in this way. They do have an economic worth of considerable magnitude, not to the parents who rear them, but to the government, employers, and the whole community (Folbre, 1994b; Chesnais, 1998; Klevmarken and Stafford, 1999; Crittenden, 2001). At a societal level, the economic benefits of children far outweigh the costs. Children are a net economic gain, but in the long term and to the whole community, not for the parents who provide the time and money involved (Klevmarken, 1999; Crittenden, 2001). 'The cumulative cost of a child to his parents in terms of time, money and energy from birth to adulthood is huge. It is a massive investment in human capital. The return on this investment is also huge, but it is not returned to the investors (the parents); rather it is absorbed by the state, pension funds and private companies' (Chesnais, 1998, p.97).

So this perspective holds that, rather than being consumption items for parents, children are in fact a product of family inputs comprising money, time and labour (England and Folbre, 1997). The economic benefits of children flow to the state and particularly to employers, while the economic effect on parents is negative. The degree to which a country accepts the public good view of children will be reflected in social policy, and rates of child-related government spending, for example on schools and child health care. Unless the costs of children are socially spread, the state, employers and the childless are free riders on the time and money inputs of parents (Folbre, 1994b; Joshi and Davies, 1999a). On this view, there is a public benefit at substantial private cost. It follows that a welfare gap between parents and the childless is an important inequity.

It also follows that it is important to establish the nature of those costs. The family inputs to children include money, but cannot be reduced to money only. They also consist of time and labour. Fully acknowledging the family contribution to raising children would require recognition of non-monetary as well as monetary inputs. As we have seen, new home economics expanded the definition of productive labour to include housework. This accords with the view, strongly put in a body of feminist literature, that housework is both laborious and productive (see for example Mitchell, 1971; Rich, 1977; Oakley, 1979; Pateman, 1988; Waring, 1988; Folbre, 1994b; Maushart, 1997). However, new home economics does not define the labour of looking after children as economically productive work, because it regards parents as the sole beneficiaries of their own children and child care as conducted for its own sake.

Challenging this has proved more difficult than redefining housework as productive labour, not least because parents gain psychic satisfaction from caring for their children. But paid work gives psychic satisfaction too. Challenging the core, but dubious, assumption of neoclassical economics that this is not the case Folbre argues that work should be seen in terms of specific activities that generate transferable benefits *extra* to any intrinsic psychic benefit derived (Folbre, 1994). Child care fits this definition because it creates transferable benefits – not only to the children themselves, but also to the parent who does not do the primary caring, and crucially and most significantly, to the whole of society. Acknowledging that children are a public good requires extending economic definitions of work to include the labour of caring for children. If caring for children is productive labour, it is important to establish the extent of the labour involved, and who performs it. This means that gender neutrality is a major barrier to full understanding of this issue.

While welcoming the acknowledgement by new home economics of the economic importance of household work, feminists take serious issue with the conceptualisation of a joint household utility function, which assumes an identity of interests and equal distribution of household resources (see for example England, 1993; Folbre, 1994b). Though theoretically gender blind, the assumption of a joint household utility function reflects a powerful masculine bias, and perpetuates the tradition of western thought that places intra-household equity beyond scrutiny (Bergmann, 1995; Nelson, 1996). The idea of a joint utility function closes rather than opens the family to scrutiny (Nelson, 1993). New home economics is imbued with essentialist assumptions about sex roles, and by assuming the family has a unified utility function, arbitrated by the (male) head of the household, it can tell us little about women's lives. Unequal inputs or outcomes by gender are not addressed. There is no recognition of potential conflict, or that any coercive authority or power may be exercised (Woolley, 1996).

In contrast, domestic bargaining theories (discussed more fully in Chapter 6) do incorporate the recognition that decisions may be contested within households, and enable gender comparisons of welfare (Woolley, 1996). Bargaining theories conceptualise intra-household decision-making about work allocation as resulting from bargaining between partners on the basis of their relative resources (Scanzoni, 1979; Manser and Brown, 1980; McElroy and Horney, 1981; Lundberg and Pollack, 1993; Brines, 1994; Molm and Cook, 1995). The basic idea is that the more resources one has that one is currently sharing but could withhold, the greater one's bargaining power (Woolley, 1996; England and Folbre, 2002). Most of these resources are personal, but bargaining theories do admit scope for acknowledging public interest in children and the importance of the prevailing social attitude to children. A social policy environment in which women can maintain an independent household following separation would also enhance their bargaining strength (McElroy, 1990; Lundberg and Pollack, 1993; Folbre, 1997).

Bargaining theories are gender neutral in that they attribute gender inequity to disparity in relative resources, not to intrinsic sex differences. They imply that improving women's personal resources, particularly income and opportunities in the workplace, will mean domestic tasks are more equally allocated by sex. However, this assumption is contested. Feminist theory has long argued that it is not only barriers met by women in the labour market and public sphere that cause non-exchange in relationships, but that it also arises directly from within the family (Delphy and Leonard, 1992). Indeed, some argue that the family is the primary site in which gender differences are created (Berk, 1985; Morehead, 2005). Also, emerging empirical evidence does not show that improving women's public opportunities leads to domestic equity.

Studies have attempted to identify factors, such as education, income, workforce participation, hours worked, social class, race, relative financial contribution and feminist values, which ameliorate the unequal division of labour (Manser and Brown, 1980; McCrate, 1987; Hochschild and Machung, 1989; Brines, 1994; Dempsey, 1997). None appears reliably predictive of male domestic contribution. While it does seem that the person with least time and most economic resources does least domestic labour and that as women do more paid work, they do less unpaid work, women do much more housework than their partners no matter how educated they are, how much they earn or how many hours they spend in paid work (Hochschild and Machung, 1989; Brines, 1994; Coltrane, 1994; Widmalm, 1998; McMahon, 1999; Greenstein, 2000; Baxter et al., 2005). The overwhelming finding across research on domestic equity is that the division of labour on gender lines is remarkably persistent.

Indeed, the persistence of a gendered division of labour has proved so impervious to variation in household circumstance that the greater challenge is not to identify the factors that result in more domestic equity, but to explain why the factors have such a minor impact (Coltrane, 2000; Dempsey, 2001). In the face of enormous changes in female behaviour, male behaviour remains remarkably unaffected (Bianchi, 2000; Baxter, 2002). Domestic labour *has* become more equal over time, but not because men now do much more, but because women do much less. Worldwide, women have substantially reduced the amount of time they spend doing housework (Dempsey, 1997; Bittman, 1998; McMahon, 1999; Bianchi et al., 2000; Gershuny, 2000; Baxter, 2002; Bianchi et al., 2006).

Some take this as an indication that sociological theories of gender norms and socialisation have more explanatory power than the ostensibly gender-neutral economic theories. Couples may avoid deviating from social norms by actively 'doing gender', that is, behaving in stereotypically gendered ways, particularly when certain norms, such as the man being the major earner, are breached (Brines, 1994; Greenstein, 2000; Bittman et al., 2003). Others argue that domestic behaviour is affected by the expectations held by others, and that people follow normative expectations to conform even when it is not to their own best advantage (West and Zimmerman, 1981; Berk, 1985; South and Spitze, 1994; Waller, 2002). Active gender strategies are used to allow behaviour to fit with gender ideologies (Hochschild, 1997).

However, while the research suggests that a gender-neutral approach does not clarify the issue, gender role assignment also seems an insufficient explanation. If people behave in accordance with gender role allocation, why have so many women embraced paid work? Why have they changed their behaviour in some ways and not in others? There has been a move towards masculinisation of female work patterns in both paid work and housework, but not in time spent caring for children. Despite widely expressed fears that child neglect would follow women's entry to the workforce, average parental time with children has actually increased over the last few decades (Bianchi, 2000; Bianchi et al., 2006). Time spent looking after children remains very high, as well as very gendered.

This suggests that a core issue in the stalled revolution whereby women's gains in the public sphere are not reflected in the private, is not gender alone but the gendered effects of parenthood and other care responsibilities. Care of children is a particularly knotty problem. Public sphere reflections of this are acknowledged. The birth of children is known to impact upon female workforce participation. Time out of the paid work force as leave or shorter hours, combined with slower promotion and career progression, mean that mothers suffer a financial penalty that leaves them considerably worse off over a lifetime than either men or childless women (Joshi and Davies, 1999b; Gray and Chapman, 2001; Breusch and Gray, 2003). Childless women have been found to enjoy more gender equity in both workforce participation and in earnings than mothers do (Williams, 2001). Williams suggests that, far more significant than the 'glass ceiling' that inhibits promotion of women beyond a certain point, there is a 'maternal wall', which means that motherhood is a greater predictor of professional disadvantage than femaleness per se.

But concentrating on public sphere indications shows only one side of the story. An inferior market position is both a cause and a result of women's subordinate position in the family (Cass, 1995; Shaver, 1995; Lake, 1999). A fuller understanding of the stalled revolution requires a complementary investigation into the private sphere that addresses both parenthood and gender issues and also differentiates between them. The discussion above suggests that the specific effects of having children are currently neither adequately theorised nor researched. Several aspects of parenthood need to be more clearly acknowledged.

First, parenthood and gender issues, though obviously related, are not the same. Mothers' needs can be neither simply fused with, nor simply opposed to, children's needs (Riley, 1983). Nor are the needs of fathers and mothers either simply opposite or simply commensurate. Women have things in common with each other; men have things in common with each other; parents have things in common with each other. What is lacking is adequate knowledge of where the differences and commonalities lie. A risk of a gender-free concept of parenting is that fatherhood, not motherhood, will be the yardstick. If new home economics' conceptualisation of children as a private leisure activity more closely reflects how men parent than how women parent, and this model of parenting is universalised, children will be affected and/or the inputs of mothers will not be fully recognised. The conceptualisation serves to obscure the specific position of mothers, and with it a clear understanding of the amount of work and time that raising children actually requires.

Also, children are of both private and public concern. Although there are potential serious social effects of failing to evenly distribute the costs of children, a lack of recognition of their social value has helped obscure their cost and who pays it. This is not a problem confined only to parents or to mothers, but has consequences for all. This book aims to ameliorate the lack of empirical research into the important social issue of parenthood and its effects, through the study of time-use data, which offer the opportunity to investigate directly how parenthood impacts on the domestic sphere, and how this impact differs by sex.

Direct quantitative research into the family is relatively rare. This is partly because of the sanctity of the private sphere, but there is also a practical reason. Good quality data have not been available. An opportunity to rectify this has come with the advent of time-use surveys that are now being regularly conducted in many countries. Surveys of time-use provide a valuable adjunct to more established statistical information regarding income, household expenditure, employment patterns, housing and demographics. Time-use data's unique contribution to research is to provide direct information about the private sphere, particularly by quantifying unpaid work, which is largely invisible to usual data collection methods. Previously, it could only be deduced indirectly, for example, through spending patterns or as the inverse of public measures such as women's work force participation. Time spent on unpaid domestic obligations is an important indicator of the intensity of familial welfare responsibilities (Esping-Andersen, 1999). Time measurements can also shed light on how the social policy environment translates into a reality in people's lives (Land, 1995; Gornick et al., 1996; Orloff, 1997; Esping-Andersen, 1999; Plantenga and Hansen, 1999). Time-use studies provide 'the most accurate current estimates of all unpaid work and family care that takes place in society, and giving an otherwise unavailable glimpse of all the things that people do' (Robinson and Godbey, 1997, pp.288–289).

The invisibility of unpaid domestic work has implications at both a personal and a social level. At a macro level, the economic value of home production is overlooked, despite an estimated value equivalent to over 60 per cent of gross domestic product (GDP) (Ironmonger, 1996). At a personal level, those who do the home production have unrecognised constraints upon them if it is invisible. Their time is consumed but undervalued. Welfare is more commonly compared in the metric of money than of time, but the problem of balancing work and family is arguably more about time constraints than about the scarcity of money resources. So time is, in itself, an important independent measure of welfare. It is particularly significant to women, especially mothers (Gershuny, 1999). It is central to issues of family and gender. Therefore, this book uses time-use data to undertake a broad-ranging and detailed analysis of the time effects of having children. A brief chapter outline follows.

Chapter Outline

In Chapter 2, I review previous attempts to quantify the cost of children, in money or in time. I explain the approach to calculating the time impacts of children used in this book, and describe the data sets, measures, and statistical methods drawn upon.

Chapter 3 shows the magnitude of the time impact of children, and how it varies with family size. I present a comparison of the daily workload of adults in households with no children and in families with different ages and numbers of children. Firstly I look at the time of couple households as a unit, and secondly I look at the time of men and women within those couple families. In a supplementary analysis, I compare the time commitments of sole parents to those of parents in couple families.

In Chapter 4, I extend and deepen the gender analysis. I show how measuring secondary (simultaneous) activity is important to recognising the full magnitude of the time gulf between parents and non-parents, between mothers and fathers, and in showing how sole parents compensate for the lack of a resident partner. I look at the ways in which mothers and fathers' parenting differs, not just in absolute terms but also relatively, by teasing out the differences in how men and women spend time with children. I reveal dimensions of double activity, responsibility and time allocation to different types of child care task that are essential to a full understanding of gendered time inputs to child care.

In Chapter 5, I examine how the use of non-parental child care affects the quantity and quality of parental time in activities with children. The use of non-parental care is not associated with the loss of an equivalent amount of parental child care time, and I conduct an enquiry into how parents shift and squeeze their time to manage this apparent paradox.

In Chapter 6, I investigate whether educational level is associated with differences in quality and quantity of time with children, in maternal time allocation to home and market production, and in the sharing of care between fathers and mothers. Education is an important variable in studying time allocation, because it is central to economic theories of time devoted to paid work, higher education is associated with liberal gender attitudes, and educated parents may be particularly influenced by theories of child development that advocate high levels of parental involvement with children.

In Chapter 7, I place the issue of the impact of children upon adult time in an international context. Comparative research provides a framework for testing the effects of alternative policy settings. I give a brief overview of how becoming a parent affects time-use in four countries (Australia, Norway, Germany and Italy), each representing a different type of social policy regime, with different approaches to economic, social and family organisation, to consider whether the time penalty of parenthood, and its distribution by gender, is influenced by social policy environment.

In Chapter 8, I conclude with a summary, overview and discussion of my findings, and canvas the social and policy implications.

Chapter 2

Approach, Data and Method

In this chapter I review the methods used in previous research to measure the financial cost of children and estimate the time costs of children. I outline my own approach, which is to compare families of different size, and men and women, on an extensive range of time-use measures. I discuss time diary methodology and describe the data sets used in this book: the Australian Bureau of Statistics Time Use Survey (1997) (ABS TUS 97), and the Multinational Time Use Study World 5.5 (MTUS). I set out the statistical methods employed, and explain the outcome and predictor variables I measure, the reasons they are of interest, and how they are presented in the book.

Calculating the Costs of Children

Children are expensive (Folbre, 2002). Families with children have higher monetary outlay and are more likely to be living in poverty than families with no dependent children (Land and Lewis, 1997). However, while it is acknowledged that children are costly, there is no agreement as to how to calculate that expense. The 'expectation of those that inquire is that the cost of a child is a fairly straightforward matter [but on the contrary] estimating the cost of children is a complex and highly imprecise exercise' (McDonald, 1990, p.19). Any set of figures reflects contestable starting assumptions and is imbued with normative values (Saunders, 1999).

Direct Financial Costs

It is difficult to calculate the cost of children even when the enquiry focuses only on the simplest measure; direct financial outlay. There is a long history of scholarship in the subject. There are several ways to attempt it. The first is to quantify expenditure on what children need, and what would therefore have to be spent to provide a minimum standard of living. The second is to calculate what is actually spent on children by their families. The third method, usually using information gathered by the previous two, is to establish how much extra income a family requires to support each extra child, and try to establish equivalence measures across family types.

In Australia, the first two methods have been used for some time. The basket of goods approach developed by Kerry Lovering, based its figures on the inflation-adjusted cost of certain expenditure items. The figures were separately calculated for low and middle-income families, and for children aged 2, 5, 8, 11 and 13 years and over, giving a weekly and yearly estimate for each category. The strength of the approach was that it calculated ideal expenditure on food, clothing, fuel, household provisions, and the cost of schooling not including fees. The weakness was that it was not comprehensive. Not included were housing, transport, school fees and uniforms, child care, medical and dental costs, gifts, pocket money or entertainment (Lovering, 1984).

The survey expenditure approach, formulated in Australia by Donald Lee in 1989, calculates actual weekly expenditure in families with children in age bands from birth to 13 years. The methodology is used internationally. For example, it forms the basis of the US Expenditures on Children by Families Report, conducted annually since 1960 by the US Department of Agriculture. A recent estimation is that depending on income level, it costs between \$US127,080 and \$US 254,400 to raise a child in the US (Lino, 2003).

While it gives a comprehensive picture of family expenditure, the weakness of the survey expenditure method is that it is unable to indicate whether parental outlays are truly appropriate to children's needs. There is no way of knowing whether the amounts spent are over-generous, adequate, or insufficient. Further, it does not really clarify the additional cost of each extra child. While it is able to quantify costs of child-specific goods such as education, for joint family goods such as housing and transport it presumes equal sharing within families, and relies on the per capita method of allocating expenses among household members equally (Lino, 2003).

The Budget Standards Unit (BSU) at the University of New South Wales attempted to improve on the above methods by bringing together elements of both. It developed a revised version of the basket of goods approach, renamed 'budget standards', which also incorporated aspects of the survey expenditure approach. Establishing the budget standard involves describing a pattern of consumption expenditure that would ensure a particular standard of living, and estimating its cost at current market prices. The method improves on Lee's and Lovering's formulae by including a wider range of commodities, as well as the proportional cost of the children's imputed share of family-owned items such as housing (Saunders, 1999).

The improvements in method mean that the information can be used to establish the marginal costs of extra family members. This can be done in two ways. The first is a simple deductive method in which the cost of a child equals the result when the budget standard of a couple with no children is deducted from the budget standard of a couple with a child (McHugh, 1999). The second is a more comprehensive method, which unpacks each budget area, and assigns them to a child's costs or a proportion to a child's costs (McHugh, 1999).

A budget standard estimates what is needed in terms of material goods and services by a particular type of family to achieve a particular standard of living in a particular place at a particular time (Saunders, 1998a). This makes comparison across situations difficult, but it is what is attempted by the third method of establishing the costs of children – household equivalence scales that indicate the relative income needs of households of different sizes and composition. Commonly expressed as ratios, they answer the question of how much income different households would need to attain the same welfare level (Nelson, 1996).

Equivalence scales

Equivalence scales estimate the relative cost of living in families of different sizes and composition. They recognise that larger families require more resources to achieve the same level of welfare. In Australia, ideas about equivalent household welfare have long informed social policy, and underpinned the 1907 Harvester judgement that allocated wage rates for men that took into account the cost of supporting a wife and three children. However, because all household members may consume some household goods simultaneously (lighting, for example) equivalence scales acknowledge there are 'economies of scale' in that each additional member of a household adds only marginally to overall expenditure (ECLAC et al., 1999). The OECD uses a parametric equivalence scale, which allocates a value of 1.0 for an adult, 0.7 for each additional adult and 0.5 for each child under 14 years (ECLAC et al., 1999). Such parametric scales are simple and are easily understood and applied.

However, because they are not based on observed behaviour, parametric equivalence scales are fairly arbitrary and rely on untested assumptions (ECLAC et al., 1999). Despite the simplicity and apparent utility of the approach, it is widely argued that equivalence scales are flawed in practice, in method and in theory (Nelson, 1992; Nelson, 1996; Lino, 2000; Bradbury, 2001). The main problem is deciding what to compare across families. An equivalence scale relies on the assumption that families that expend the same proportion of their income on a specified basket of goods and services have the same standard of living. The central problem is how to ensure that the standard of living that is to form the basis of comparison is constant across different areas of the budget and across different households (Saunders, 1998b).

Historically, there have been several attempts to identify such a measure. Engel selected the criterion of food expenditure on children in 1875, suggesting that families in which the proportion of household spending on food was the same could be said to have the same level of overall welfare, and others followed this approach (Seneca and Taussig, 1971; Espenshade, 1984). Critics, however, suggest that this method overestimates the cost of children. It assumes they require the same share of family expenditure for every commodity as they do for food, and does not acknowledge the greater economies of scale implicit in a child's proportion of, for instance, housing costs. Nor does it take into account that the presence of children can be associated with a reduction in certain types of food expenditure such as restaurant meals (ECLAC et al., 1999; Ekert-Jaffe et al., 2000).

In an attempt to overcome these problems Rothbarth (1943) suggested calculating expenditure on 'adult goods'. The addition of children to a household reduces the amount of money available to each person, and so two households that have the same amount of money left to spend on goods consumed by adults only have the same level of welfare. Under this conceptualisation, the cost of a child is seen as the amount of money required to maintain the same expenditure on adult goods as before the child was born (Bradbury, 1992; Nelson, 1992; ECLAC et al., 1999).

To make comparisons, all equivalence scales contain the assumption that preferences will remain stable, but the Rothbarth method requires the further assumption that the presence of children has only income effects on parental consumption (Gronau, 1991). That is, changes in expenditure on parental goods will result only from changes in the amount of money available. Consistent with neoclassical economic theory, it presumes that tastes do not change and does not allow for the possibility that adults may change their spending priorities upon parenthood (Lino, 2000). The Rothbarth model also requires the assumption that adult goods can be kept separate and not shared by all family members. This raised a practical difficulty, which is that it is hard to isolate goods consumed by adults only. There was a related theoretical concern. The assumption that goods are private seems to deny the presence of household economies of scale (Nelson, 1992; Bradbury, 1995). Some assumption of 'demographic separability' is a necessary part of all attempts to create consumer equivalence scales (Gronau, 1988), but it is implausible to regard much household consumption as separable (Nelson, 1992).

In the light of these inadequacies, Barten (1964) further modified the equivalence scale method, but he assumed that all households would consume the same goods and did not allow for expenditure on specifically child goods by families with children. Gorman (1976) addressed this by adding a number of child-related fixed costs. The resulting Barten-Gorman model retains the assumption of stable adult preferences, but allows for the presence of family goods and for the possibility that the addition of extra family members will cause intra-household prices to vary. That is, new members bring direct demand, and also the indirect effect of making certain goods relatively more expensive (ECLAC et al., 1999). The model therefore allows for substitution between commodities (Browning, 1992). However, it is not possible to establish intra-household allocation. The model assumes individuals in the household are identical, and that the household welfare is symmetric. This implies, for example, that there is no difference in household utility according to whether cigarettes or milk is consumed. Indeed, it implies that the former is more likely as cigarettes become relatively cheaper with additional children (Nelson, 1992). Thus, the model contemplates implausibly high degrees of *quasi*-price substitution, and cannot really explain how or why adults redirect utility to children (Bradbury, 1995; Nelson, 1996).

Identification of this practical weakness in the Barten-Gorman model jelled with a conceptual development in economics that regards the happiness of the adults in a family as the true measure of household utility. The concept of equivalence scales as a basis for welfare comparisons sits uneasily with neoclassical economic theory that emphasises the role of individual preferences in consumption patterns and the impossibility of comparing subjective utility (England, 1993; Nelson, 1996). As discussed in Chapter 1, within neoclassical economic theory children have come to be conceptualised as equivalent to consumer goods for parents. This means many economists view equivalence scales as a meaningless basis for inter-household welfare comparisons, because children are a benefit to the parents, for which compensation is unnecessary (Bergmann, 1995; Nelson, 1996; Bradbury, 2001). Equivalence scales that concentrate on the material costs of children without factoring in the assumed psychic benefit to their parents have come to be regarded as 'conditional equivalence scales' which should not be used to draw conclusions about comparative household welfare (Pollack and Wales, 1979; Blundell and Lewbel, 1991). Parents, through exercising their reproductive choice, can be presumed to have decided the costs of children are outweighed by the psychic benefits (Bradbury, 2001).

So practical enquiry into standard of living across different sizes of household fell out of intellectual fashion in favour of assuming increased expenditure following parenthood is anticipated and accepted. The individualistic theoretical thrust of economics sidelined questions about how parenthood, and having extra children, affects welfare. Comparisons of welfare are not meaningful within that framework (England, 1993). Implicit is a rejection of the view that children have a public value. As canvassed above, equivalence scales also have practical problems in that no existing method fully manages to reach the aim of taking account of economies of scale in household size, price-like substitution effects and intra-household allocation of consumption. The combination of theoretical challenge and practical deficiency were almost sufficient to discourage the use of equivalence scales entirely (Lino, 2000).

However, there is resistance to allowing the current focus on adult subjective utility to assign the method to oblivion. Sen acknowledges that undertaking sympathetic tasks, such as those involved in caring for children, may make a person feel better, but argues that there is still a case for objective standard of living comparison (Sen, 1998). Nelson (1996) points out that while concentrating on subjective utility may advance economic theory, it is unhelpful empirically. She suggests that rather than abandoning the method, it would be better to enhance it by enlarging the measures of welfare, singling out time-use as one example (Nelson, 1996). Comparing welfare solely on the basis of monetary expenditure misses the important point that children also place extremely heavy demands on parental time. This is the issue I address in this book. I calculate the time impact of children in order to better inform standard of living comparisons between and within households.

Before describing my own approach to estimating the time cost of children, I review the methods employed in previous research. How to establish the time cost of children is, like establishing their financial cost, both difficult and contested.

Time Costs

Indirect time costs: Opportunity costs

Most research into how time inputs to childrearing affects parental time allocation has concentrated on the flow-on effects on workforce participation, in particular the opportunity cost of foregone maternal wages. This body of research shows that this indirect subsidy dwarfs direct financial outlay on children. The major part of the costs of children is the opportunity cost of forgoing waged labour to care for them (Apps and Rees, 2000; Breusch and Gray, 2003; Browning and Waldfogel, 1998a; Joshi, 1998; Joshi and Davies, 1999b; Davies et al., 2000; Lechene, 2003). Estimates suggest that mothers lose about 60% of lifetime earnings compared to women who have no children. The opportunity cost approach allows a dollar value to be put on motherhood. For example, an Australian study found the effect of a first child to be a cumulative loss of earning over a woman's life course of approximately (in 1997 dollars) $A435,000 (Beggs and Chapman, 1988). The economic penalty for devoting time to children goes beyond immediate wages unearned. Time out of the workforce has long-term downward effects on employability and superannuation benefits (Waldfogel, 1998b; Joshi, 1998; Davies et al., 2000). There is no corresponding loss for fathers. Men who have children have been found to earn more over their lifetime than childless men (Joshi and Davies, 1999b; Davies et al., 2000).

Quantifying the opportunity cost of mothers' foregone earnings due to time spent out of the paid workforce is a very valuable addition to quantifying direct financial expenditure on children. However, it has limitations. Equating the value of time spent with children with the amount of wages foregone has the drawback of being related to the market value of work skills rather than parenting skills (Horrigan et al., 1999). This means, for example, that a professional woman's time with children would be valued more highly than that of a factory worker, when the latter may be as good or better at child care itself. At the extreme, women who do not have paid employment before motherhood would have their mothering time valued at zero (Bojer, 2006). Also, its market-focused approach provides only a partial account of the time demands associated with raising children (Joshi and Davies, 1999a; Klevmarken, 1999; Apps and Rees, 2000; Folbre et al., 2005). While lost earnings are important, so is the actual time expended. Time with children is added to time in paid work as well as substituted for it (Joshi, 1998). Calculating opportunity cost ignores time spent with children that is withdrawn from activities other than paid work, such as leisure, sleep, personal care and social activities. The opportunity costs approach does not give a full accounting of the time impact of children.

Direct time costs

There have been studies that quantify direct parental time inputs to children (see for example Bianchi, 2000; Craig, 2002b; Sayer et al., 2004). This method is also problematic because of disagreement about what should be measured and how it should be valued. Currently, there is no agreed upon satisfactory method of quantifying direct time investment in children (Budig and Folbre, 2004). The usual approach is to calculate the amount of time spent in active care of children as a main or 'primary' activity. But because a very high proportion of child care is done at the same time as other activities, this gives a radical underestimation of parental time investment in children (Ironmonger, 1996).

Counting child care only when done as a main activity ignores time supervising children, or engaged in other activities but available to be called upon. This type of 'passive' supervision is a large part of overall care, as young children cannot be left alone, and because sensitive, observant monitoring of play is a vital and demanding part of child care (Leach et al., 2005). To ignore this aspect of the time demand of children obscures constraint upon carers and is a serious under-representation of total time inputs. However, even including secondary activity will underestimate the full parental time input to children since it excludes time away from children that is spent performing essential duties such as planning activities, shopping, cleaning up, making arrangements, or being available to be summoned.

On one view, all parental time, even that spent sleeping or not in the child's company, should be counted in estimates of care (Budig and Folbre, 2004).

The measurement problem is intertwined with the valuation problem. How to value time spent with children remains an open question. Unpaid work can theoretically be valued by outputs, such as placing a monetary value on a homemade dress, shed or meal, equivalent to what would be paid commercially. However, this is problematic when activities create no visible product, as with child care. A simpler approach is to place an economic value on the time inputs. So value can be deduced by imputing a wage to the time spent in child-raising activities (Horrigan et al., 1999). But the difficulty here is to decide which child-raising activities justify a wage. This is informed by the issue of whether children are a public good or a private pleasure, though this is not always articulated. If caring for children is conceived as a consumption item or leisure activity for the parents, no wage would be appropriate. On the other hand, if they are performing a socially useful service, putative wages may be justified.

Researchers have attempted different solutions to this problem. Juster suggests an approach that implicitly recognises the public value of children, but places limits on it. He suggests a wage value be placed on time spent in helping children make or do things, in teaching them new skills and providing for their healthcare, plus 0.5 of the time spent caring for, reading to, talking to, playing with and being a chauffeur to one's children (Juster, in Klevmarken, 1999). This approach conforms to Ironmonger's definition of work as activity that benefits others (Ironmonger, 1996), as Juster focuses on those inputs that could be seen to contribute to the child's human capital, and therefore to have a social benefit. He excludes activities that provide pleasure or 'process benefits' to the parent (Klevmarken, 1999).

England and Folbre take the much broader view that *all* parental time with children is material to producing a useful and well-rounded citizen (England and Folbre, 1997). Even if pleasurable to the parent, the product of care is of primarily social value. In this sense it is not different from paid work, which, though sometimes pleasurable, is economically regarded as primarily to produce a good or service. At different wage rates reflecting the degree of involvement, Folbre costs all parental time, even that spent sleeping or not in the child's company. Adding her result to the estimate of the cost of children reached by the US Department of Agriculture yields a figure four times higher than when financial outlay alone is counted (Folbre, 2004). This is a further indication of the enormous value in the non-market sector of the economy, which Ironmonger has estimated to be equivalent to 60% of GDP (Ironmonger, 1996). If home production is excluded from national accounting, it is impossible to see whether increases in the market sector are genuine productivity gains, or simply a transfer of activity from the unpaid sector to the market economy (Gershuny, 2000). So Folbre's analysis is a powerful indicator of the magnitude of the social value of time devoted to children. However, it is less helpful in clarifying standard of living and welfare effects resulting from parenthood upon mothers and fathers themselves.

Each of the methods of valuing parental time outlined above offers increased understanding of the issue, but none is definitive. Currently, there is no recognised way of calculating parents' time input to children that acknowledges both market and non-market effects, and that focuses on the impacts upon parents themselves. My own approach is to use a time measure that foregrounds the impact of children in the home. It does not put a monetary value on time devoted to children, but retains acknowledgement of indirect effects from and on market participation. Translating every effect of children into money may divert attention from the fact that time scarcity may be most appropriately alleviated with time assistance, not financial compensation. Also, monetary measures are less gender-sensitive than measures of time. The presence of children brings with it a requirement to perform caring duties, not just extra money outlays. The current problems of balancing work and family demands that result from becoming a parent are arguably as usefully regarded in terms of time constraint as they are in terms of money scarcity.

Marginal time costs

For the first part of the analysis in this book, I loosely adapt the equivalence scale approach to studying the comparative welfare of adults living in different family configurations. I compare the daily workload of families with different numbers and ages of children, to calculate the marginal differences in workload that are associated with each extra child. Daily workload is an aspect of standard of living that is missing from expenditure equivalence scales. Yet it is an important indicator of welfare. People who must work long days are disadvantaged compared to those who do not.

Calculating daily workload puts the caring and domestic labour responsibilities associated with children at the centre of analysis, while also acknowledging market work, but leaves the calculation in the metric of time. However, how that time is composed is also important. As a starting point, I calculate three progressively encompassing measures of the impact of children on parental activities, each of which is a subset of the next. In order of increasing inclusiveness these are: child care; unpaid work (domestic labour plus child care); total work (paid employment plus domestic labour plus child care), and total work (paid employment plus domestic labour plus child care). Each successive measure captures a wider impact of children. The measures acknowledge financial input to children through paid work as well as direct input to children through unpaid work. Unpaid household work escapes economic measure, and paid work, necessary to the maintenance of children, is not counted in quantification of direct care of children. Both are important, and this method allows both to be represented in the analysis. Total (market and non-market) work time captures the broad time impact of parenthood including any compensating adjustments in the household supply of market work. However, not least because it is not remunerated, it is also important to know the magnitude and gender distribution of unpaid work, and how it is divided into housework and child care. Further, it is necessary to investigate how time with children is actually spent, including exploration by gender of specific child care activities performed, multitasking and double activity and who is present. Specific measures are detailed after the data set descriptions below.

Data

The analysis uses data from the Australian Bureau of Statistics (ABS) Time-Use Survey (TUS) (1997) and the Multinational Time Use Study (MTUS) World 5.5.

Australian Bureau of Statistics Time-Use Survey (1997)

The TUS (1997) is the most recent in a regular series of cross-sectional time use surveys conducted by the ABS. The TUS randomly samples over 4,000 households, requiring each person aged 15 years or older resident in that household to record, at 5-minute intervals, all their activities over 2 days. To capture seasonal variation, the time diaries are collected at four separate periods over the calendar year. Activities are divided into 10 broad categories: personal care, labour force, education, domestic, child care, purchasing, voluntary work and care, social and community interaction, recreation and leisure. The survey provides accurate information about the start and finish time of activities, simultaneous activities, the location of activities and the company present. Respondents note their main or 'primary' activity, report what else they were doing at the same time (secondary activity), note the location and who else was present during the activity. The surveys collect extensive demographic data, and the average number of episodes per day (over 30) garnered by the TUS indicates higher than usual data quality. There is low non-response distortion because under Australian law, cooperation with the ABS is mandatory and the rate for fully or partly responding households is 90%. Within this, over 84% of persons were fully responding. The final sample of the survey was 14,315 diary days (ABS, 1998).

For the main part of this analysis, the sample is restricted to households in which the only adults are either childless couples or couples co-resident with at least one child under 12 years old (n = 1,210 households, n = 4,274 person-days). This restriction is intended to provide the clearest point of comparison on the basis of parenthood and number of children, and therefore excluded from the sample are households with older children or other co-resident adults who could provide intra-household child care in addition to parents. The age range of the adults is restricted to those between 25 and 54 years old. This simplifies the analysis and interpretation by removing younger full-time students and 'early' retirees from the investigation, and concentrating on the age range most crucial to establishing a family and developing a career.

Where appropriate (for example to investigate the child care inputs of parents by sex), the sample is further limited to parent households only (n = 705 households, n = 2,926 person-days). Where non-parental child care is the focus of analysis, the sample is of individual parents in households with a youngest child under 5 years of age (n = 1,690 person-days). In one instance the sample is extended. Finding a point of comparison for sole parents is not straightforward. It is not obvious whether sole-parent households should be compared to single-person households or to households with two parents. Consequently, lone parents and single people were added to the sample in those parts of the analysis intended to investigate the time impacts of sole parenthood (n = 6,035 person-days). When child care was the focus of comparison, this sample was restricted to individual sole and couple mothers (n = 1,708 person-days).

Most of the analyses are conducted at both household and individual level. For the household level investigation, each household is represented by one joint diary day. Otherwise, the count is person-days. This does mean that the observations are not entirely independent, but because the pattern of activities varies by day even for the same individual, retaining both day records of each respondent gives a fuller picture of their time-use than amalgamating the days into one.

Multinational Time Use Study (MTUS) World 5.5

The MTUS is a data set in which over 50 time use surveys in 19 countries are harmonised, to facilitate cross-national comparisons. It began in the late 1980s, is run from Oxford University by Professor Jonathan Gershuny and Dr Kimberly Fisher, and is being constantly updated as new national surveys are completed. From the MTUS I created a sub-sample of individuals in couple-headed families in Australia (survey year 1992 n = 5,905), Germany (survey year 1992 n = 7,761), Italy (survey year 1989 n = 13,457), and Norway (survey year 1990 n = 2,644).

Method

Measures

The measures analysed are hours a day spent in various activities. The first group is paid and unpaid work activities. Specifically, they are

1. Paid work (ABS codes 200–299): employment related and training activities – main job; other job; unpaid work in family business or farm; work breaks; job search; attendance at educational courses; job related training; homework/ study/research; breaks at place of education; communication and travel associated with these activities.
2. Domestic labour (ABS codes 400–499): housework; food or drink preparation and meal clean-up; laundry, ironing and clothes care; tidying, dusting, scrubbing and vacuuming; paying bills and household management; lawn, yard, pool and pet care; home maintenance and pet care; shopping for goods and services; communication and travel associated with these activities.
3. Child care (ABS codes 500–599): child care is an extremely heterogeneous activity. Parental interaction with children ranges from passive, supervisory care to highly demanding forms of interaction that have important consequences for the development of their personal capabilities. Most studies simply tally child care in terms of total minutes of parental time. But the activity codes available in the TUS can be used to create a typology of categories ranging from the most to the least intense interactions. I define four categories as follows:
    a. Interactive (talk based) child care (ABS activity codes 521 and 531): Face-to-face parent-child interaction in activities teaching, helping children learn, reading, telling stories, playing games, listening to children,

talking with and reprimanding children. These activities are critical for the development of children's linguistic, cognitive, and social capacities (Brooks-Gunn, Han and Waldfogel, 2002).

b. Physical child care (ABS activity codes 511 and 512): Face-to-face parent-child interaction that revolves around physical care of children. Feeding, bathing, dressing, putting children to sleep, carrying, holding, cuddling, hugging, soothing. This is nurturing care that fosters security and emotional wellbeing in children (Leach, 1977).

c. Travel and communication (ABS activity codes 57 and 58): Travel can be associated with transportation to school, visits, sports training, music and ballet lessons, parents and teacher nights. Travel time includes time spent waiting, and meeting trains or buses. Communication (in person, by telephone or written) includes discussions with a spouse, other family members, friends, teachers and child workers when the conversation is about the child.

d. Minding children (passive child care) (ABS activity code 54): supervising games and recreational activities such as swimming, being an adult presence for children to turn to, maintaining a safe environment, monitoring children playing outside the home, keeping an eye on sleeping children.

The second group of measures is non-work activities, which are reduced following parenthood. These are

4. Personal care (ABS activity codes 100–199): sleeping; sleeplessness; personal hygiene (bathing, dressing, grooming); health care; eating/drinking; associated communication; associated travel.

5. Recreation and leisure (ABS activity codes 900–999): sport and outdoor activities; games, hobbies, crafts; reading; audio/visual media; attendance at recreational courses (excluding school and university); other free time; associated communication; associated travel.

The variables are calculated in three distinct metrics: 1) as a main (primary) activity; 2) as either a main or simultaneous (primary and secondary) activity; and 3) as time when the activity was being done while in the company of certain other people. To avoid double counting in the variables including secondary activity, when any one activity was recorded as both a secondary and a primary activity I count the time span only once. Because the focus of this study is on workload, I exclude time in which child care is recorded as a secondary activity to sleep.

I also use the secondary activity coding in another way. From time in which leisure is coded as primary activity I deduct any time during which child care is also coded as secondary activity to create the variable

6. Leisure without simultaneous child care duties.

Finally, as mentioned above, the TUS asks people to record whom they were with at all times during the diary days. This means it is possible, in addition to the recorded information on child care as a discrete activity, to measure the time parents are in the company of their children, and who else is present. Using this information, I calculate measures of hours a day in the activities above while alone, while in the company of other adults (including one's spouse) or while in the company of children only. As this type of variable I created

7. All time in the company of children, and subsets of this variable:
    a. Time alone in the company of children (no other adult present).
    b. Time performing child care activities while in sole charge of children (no other adult present).
8. Leisure with no children present.

This extensive list of dependent variables is aimed at giving as comprehensive an account of the magnitude and composition of daily workload, and of the quality and quantity of care provided by parents, by gender, as is possible with the available TUS data. The reasons why these measures are of interest are, where appropriate, set out more fully in the chapters that present the relevant findings. The extensive and intensive analysis possible with the TUS could not be replicated with the MTUS data. For the cross-national analysis, the measures investigated are time spent in total work (paid and unpaid), and time spent in unpaid work as primary activities only.

Limitations

The data have a number of limitations. They are cross-sectional, so it is possible to compare demographic groups at one point in time, but not to follow individuals moving through the life course. The TUS data are collected only from household members over 15, which means that child care is seen only from the parents' perspective, as there is no diary information directly from children. Moreover, in families with more than one child, it is not possible to determine how parental child care time is allocated to each child. The relevant variables record only the total parental care time and not the time for each individual child. The data cannot provide answers to some of the questions that could be of interest in relation to caring for children. For instance, the data do not capture planning time, or the pervasive responsibility of being a parent. Similarly, the survey only records behaviour, and does not indicate how respondents feel about what they do. The MTUS has more serious data limitations, which are discussed along with limitations of cross-national study in Chapter 7 (p.204).

Multivariate Analysis

The statistical method used is Ordinary Least Squares (OLS)[1] multivariate regression modelling. Multiple regression aims to show the relationship between

---

1   Because time use data almost always contains a high number of zero observations, some econometric analyses of time use have used Tobit regression modelling. However, time-use specialists prefer OLS to Tobit (Jay Stewart, 2006; Jonathan Gershuny and Muriel Egerton, 2006) because time use zero observations do not arise from the data being truncated

several independent or predictor variables and a dependent or criterion variable so the effect of the variables of interest, net of other possible influences, might be estimated. A limitation of the method is that there will almost always be residual unpredicted or unexplained variation, and the relationships found are not proof of causality, but of association. However, while few models perfectly fit the variance in human behaviour, the aim is to make the range of factors that may also influence the outcome as comprehensive as possible. In this study, the intention is to isolate the effects of a range of variables on adult time, measured as described above, holding constant ('controlling for') a range of other factors known or thought to also influence adult time. There are a number of independent variables that are addressed as the major focus of interest in separate chapters of this book. The model varies slightly from chapter to chapter. For instance, some variables are amalgamated into broader categories when they are not the central focus of analysis in the chapter, and the sample is sometimes limited to a particular population such as families with pre-school children. The limited MTUS data requires that the model be adapted for the cross-national chapter. Where appropriate to the subject of the chapter, regression analyses are conducted in two parts. The first set of analyses is run on households to see how the independent variable of interest impacts on the outcome variable of the family unit as a whole. Then the regressions are run for men and women separately to establish how the effects differ by gender.

Independent Variables

Family configuration is the focus of interest in Chapter 3. It is represented by a series of dummy variables combining the number and age of children as follows: no children (omitted category); one child aged 0–2 (yes = 1); one child aged 3–4 (yes = 1); one child aged 5–11 (yes = 1); two children youngest aged 0–2 (yes = 1); two children youngest aged 3–4 (yes = 1); two children youngest aged 5–11 (yes = 1); three or more children youngest aged 0–2 (yes = 1), three or more children youngest aged 3–4 (yes = 1); and three or more children youngest aged 5–11 (yes = 1). For subsequent chapters the age of the youngest child is amalgamated into two dummy categories, 'under 5' or '5–11'. Where the analysis focuses on families with children, one child is the default category, and the number of children is entered as a two dummy variables: 'two children' and 'three or more children'.

Weekly hours of *non-parental child care* used is included as a continuous variable in Chapters 3–6. It is the major focus of interest in Chapter 5, which addresses the issue of substitution for parental care. Entered into the model for Chapter 5 only is the type of non-parental child care used. Also for Chapter 5, the sample is limited to families with pre-school children, that is, under 5 years old. *Sole parenthood*

---

or censored, the circumstances in which Tobit models are appropriate (Jeffrey M. Wooldridge, 2003; William Green, 2003). Since in time use zero observations are as likely to result from non-participation as from data limitations, Tobit models that impute values will produce biased results. Recent intensive methodological investigation of this issue by time-use specialists has established that OLS is statistically more appropriate than Tobit (Jay Stewart, 2006; Jonathan Gershuny and Muriel Egerton, 2006; Jude Brown and Peter Dunn, 2006).

is entered as a dummy variable (yes = 1) in analyses in Chapters 3–6. Because of the low numbers of custodial sole fathers in the sample (4), this analysis cannot be done for men. Sole parenthood is not included as a predictor variable in the analysis in Chapter 3 because, as discussed above, finding a point of comparison for sole parents is not straightforward. It is not included in the MTUS analysis in Chapter 7.

The effect of *education* on workload and time with children is the subject of Chapter 6. For the household level analysis the education categories are: neither partner has qualifications (omitted category); one partner has vocational qualifications, other partner has no qualifications (yes = 1); both partners have vocational qualifications (yes = 1); one partner is university educated, the other partner has vocational qualifications, or no qualifications (yes = 1); both partners are university educated (yes = 1). The categories for the individual level analysis are: no qualifications (omitted category); basic vocational qualifications (yes = 1); skilled vocational qualifications (yes = 1); university diploma (yes = 1); bachelor degree (yes = 1); postgraduate qualifications (yes = 1). For the smaller sample in Chapter 5 the variables are collapsed into broader categories; 'has university qualifications' and 'has vocational qualifications'. In Chapter 7, using the MTUS data, 'above secondary education' is entered as a dummy variable (yes = 1). In all cases the reference category is no tertiary education.

For the cross-national enquiry, the independent variables of interest are *nationality*, *sex*, *family configuration* and interactions between them. The models and method are explained in more detail in Chapter 7, because they are used only in that section.

Control Variables

Entered into the models as control variables are the *age* of each spouse, split into 3 dummy variables – 25–34 (yes = 1), 35–44 (the omitted category) and 45–54 (yes = 1). *Weekly household income* in dollars per week (range \$A0–\$A2300) is entered as a continuous variable. In Chapter 8, using the MTUS data, I use two dummy variables 'lowest quartile of income range' (yes = 1) and 'highest quartile of income range' (yes = 1). The reference category is the middle 50% income range. In the individual TUS analyses weekly hours of *paid work* is entered as a continuous variable (range 0–69). The exception is when paid work time is the dependent variable; here, 'weekly hours of paid work' is omitted from the model. In the household level analysis, a series of dummy variables representing household workforce participation: both partners work full time (yes = 1), husband works full time, wife not employed (yes = 1) husband not in full-time work (yes = 1) are entered in the model. The modal arrangement in Australia (husband works full time, wife works part time) is the omitted category. In Chapter 8, using the MTUS data, I use two dummy variables 'works part-time' (yes = 1) and 'not employed' (yes = 1). The reference category is employed full-time. Also entered in the model used in Chapters 3–6 is the usual hours spent by the respondent's spouse in paid employment (range 0–60 +). In Chapter 7, using the MTUS data two dummy variables are entered; 'spouse works part-time' (yes = 1) and 'spouse is unemployed or not in the labour force' (yes = 1). The reference category is spouse working full-time.

Since time diary data are daily and the pattern of activities varies by *day of the week* even for the same individual, there is a dummy variable for Saturday (yes = 1) and for Sunday (yes = 1). The reference category is any weekday. In the household regressions, the dummies are: 'both diary days weekend'; 'diary from Saturday and weekday'; 'diary from Sunday and weekday'. The reference category is both diary days falling on a weekday. A dummy variable (yes = 1) controls for the presence of a *disabled household member*, which could be associated with unpaid work that is unrelated to the presence of children.

Unless otherwise stated, the reference category households are childless couples aged 35–44, with no post-school education, men work full time and women work part time, use no extra-household child care, on a weekday. Reference category individuals are aged 35–44, with no children, no post-school education, men work full time and women work part time, use no extra-household child care, on a weekday. The model specifications are summarised in Table A1 in the appendix.

Conclusion

In this chapter I reviewed the ways in which the cost of children has been estimated in the past, and argued that the time effects of parenthood have not yet been adequately explored. I outlined the ways this book addresses this research gap. It loosely applies a marginal costs approach, investigates an extensive and original range of measures of the impact of children upon their parents' time and explores the differential effects of family configuration, non-parental care, education levels, lone parenthood and welfare regime. I described the data drawn upon, the measures used, and the model employed in the multivariate analyses. In the chapter that follows, I present the results of the analysis of the difference in daily time commitment between parents and non-parents, and between families with different numbers and ages of children.

Chapter 3

The Time Penalty of Parenthood

In this chapter, I present findings on the difference in daily time commitment between parents and non-parents, and between families with different numbers and ages of children. The results are in three parts. The first is couple households, the second men and women separately, and the third sole parents.

Time Allocation by Couples Jointly

Child care

A couple begins to allocate time to child care when a first child is born. This is a new time requirement, analogous to direct monetary outlay on child-specific goods such as cribs, diapers and toys not required by childless households. The initial time demand is very large. From a standing start of zero time spent in child care, the birth of a first child brings a daily household time allocation of four and a half hours. (Recall that all the analysis in this chapter counts child care as a primary activity only, so even this reckoning is an underestimation. This issue is discussed more fully in Chapter 4.)

It is the impact of an infant which has the most pronounced effect on household time allocated to child care, which follows a downward trajectory as the children mature (see Figure 3.1). When there are no siblings, there is steady decrease of household child care time. The initial allocation of over four and a half hours a day when there is an infant diminishes to just over three hours a day when there is a 3–4-year-old to one and a half hours when there is a primary school child. The continuous decline after age 5 shows the strong influence of the child's entry to school, and the importance of extra-household institutions in lowering the time demands of parenthood. The timing of this fall in adults' time spent in child care may differ cross-nationally. In Australia, the use of day care for under-fives is not as widespread as elsewhere, and school entry is a rite of passage that has profound effects on parental time-use. In countries with more established institutional child care, this drop may occur earlier in children's lives.

Having subsequent children is associated with only marginal increases in time devoted to child care. Families with an infant (0–2 years old) and one additional child allocate about an hour a day more time to child care than families with just one infant. So rather than a second child occasioning a 100 per cent increase on the child care time requirement of the first, the proportional increase is 22 per cent.

The proportion of time added by a second child varies with the age of the youngest. When the youngest child is 3–4, household child care time is 27 per cent

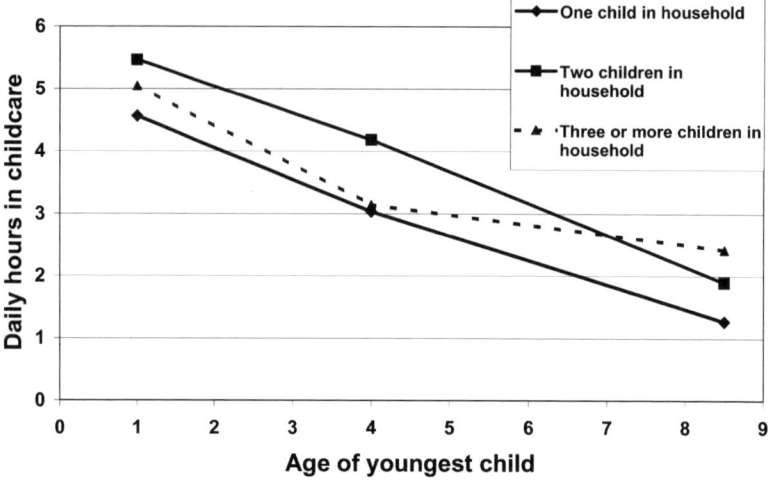

Figure 3.1   Predicted hours a day spent by couples jointly in child care (as a primary activity) by number of children and age of the youngest child

Source: This figure shows fitted values from Table A2.

higher when there is a second child. With a youngest child aged 5–11, if there are two children in the family there is about half an hour more household child care time, a proportional increase of 42 per cent. So although household child care time drops substantially as children age, with families of two, the marginal differences are more pronounced as the children mature. This means that the time cost of additional children, though less in absolute terms, becomes proportionately higher the older the children are.

This holds true when there are three or more children in a family. The marginal differences are even less in larger families. The effect is so powerful that, unexpectedly, there is less time allocated to child care in a family with a youngest child under school age in a family with three or more children than in a family with two children. This puzzling relationship does not exist when the youngest child is over school entry age – when the youngest child is aged 5–11 there is about half an hour more household child care time associated with each additional child. With this age group, a second child brings a proportional increase of 50 per cent child care time over one child, and a third child brings a proportional increase of child care time over two children.

With the data to hand, it is only possible to speculate on why there is less child care performed in larger families (three or more children) with preschoolers than in smaller families. It may reflect a selection effect. Parents may choose to have small families in order to invest more time in each child, and couples that prefer to have three or more children may have different attitudes. Alternatively, families with extra children may become more efficient at doing child care tasks, and third, older children may be contributing to child care of their younger siblings. Fourth, it may be because children who are the youngest of three are more commonly the

Table 3.1   Coefficients of hours a day spent by couples jointly in subcategories of child care by family configuration

	Physical Care	Interactive Care	Travel	Passive Care
Youngest 0-2				
Constant term	2.02 ***	1.09 ***	0.34 ***	0.82 ***
2 children	0.68 **	0.09	0.23 **	-0.10
3+ children	0.28	-0.37 ***	0.52 ***	-0.01
Youngest 3-4				
1 child	-1.33 ***	-0.08	0.38 **	-0.47 *
2 children	-0.57 *	-0.36 *	0.55 ***	-0.05
3+ children	-1.36 ***	-0.62 ***	0.59 ***	-0.06
Youngest 5-11				
1 child	-2.03 ***	-0.80 ***	0.17	-0.64 ***
2 children	-1.83 ***	-0.83 ***	0.42 ***	-0.49 ***
3+ children	-1.58 ***	-0.79 ***	0.58 ***	-0.52 ***

* P-value<0.05 ** P-value<0.01 *** P-value<0.001
Source: Authors calculations of ABS TUS 1997.The figures in this table are drawn from Table A4.

only pre-schooler in their family than children who were the youngest of two. In the sample, 71 per cent of third children have next-older siblings who are over school age, compared to second children, of whom 57 per cent have next-older siblings at school. As Figure 3.1 implies, school entry is associated with a significant drop in parental time commitment, and therefore a family in which the youngest child is the only pre-schooler will spend less time in child care tasks than families with two pre-schoolers.

Some of the explanation also lies in how parents actually spend time with children in families of different configuration. By disaggregating child care into four distinct categories – 1) physical and emotional care of children, 2) talking to/playing with/ reading to/reprimanding children (interactive or talk-based care), 3) passive care of (minding) children, and 4) travel and communication associated with children[1] – and limiting the sample to households with children, I show which sub-categories of child care time have additional parental input with extra children, and which remain at a similar level despite variation in family size (see Table 3.1).

This more detailed analysis of parents' time spent in child care by family size shows that the time saving in larger families comes mostly from interactive, talk-based care. When the youngest child is below school age, time in this type of activity is much lower in families with three or more children than in smaller families (see Table 3.1). Parents in larger families with a youngest aged 3–4 years spend less time in verbal interaction with their children when their child has the company of a sibling, and even less time verbally interacting with their children when each child has the opportunity to relate to multiple siblings. Regardless of family size, time devoted to interactive child care decreases as the youngest child matures.

---

1   As described in Chapter 2.

Parents' time devoted to minding and to transporting children is higher in households with more than one child than where there is only one child. The increase associated with the presence of a third child seems to be relatively small and in the case of passive child-minding very small indeed. The patterns of increase with the age of the youngest child are not consistent across family configuration, but broadly speaking time spent transporting children rises with age once there is more than one child present. It is the only child care sub-category for which this occurs.

Irrespective of family size, time devoted to physical care of pre-schoolers declines very steeply as the child matures. Time in this child care activity exhibits the same, seemingly anomalous, pattern as parents' child care time in aggregate. Compared with parents' physical care time spent with an only child, the presence of a second child is associated with a significant increase. However, the arrival of a third child actually reverses the sequence, so the parents' time spent in the physical care of three children is barely different from the time spent in this type of activity with one child. The finding is strongest when the youngest is below school age.

The child care timesaving that was particularly noted in families with three or more children is mainly made by reductions in household time spent in physical child care, and in interactive, talk-based child care.

Unpaid work

I now turn to the effect of family configuration on unpaid work. While the analysis of direct child care time shows a great deal about the impact of parenthood on household time, it does not give a full picture, partly because in addition to the direct, strictly child care activities coded in time diaries as 'childcare', children create more housework. For example, children's meals are prepared along with adult meals, keeping the house tidy involves clearing away children's toys, and doing the laundry involves washing children's clothes. But these activities are classified not as child care but as domestic labour, and therefore, calculating time spent in child care alone underestimates the time cost of children. To capture a fuller picture, it is necessary to analyse household time allocation to all unpaid work.

The presence of a first child has a major impact upon household time in unpaid work as a primary activity (see Figure 3.2). A household (on the mean income, in which both adults are aged 35–44 and have no post-school qualifications, in which there is no disabled family member) with no children allocates nearly six hours a day to unpaid work. The presence of one child aged 0–2 nearly doubles that, to a total of 10.9 hours a day. This is an addition of household time in unpaid work of five hours a day. As shown above, child care accounts for four of those hours. So, as with financial outlay that goes beyond direct expenditure on child-specific goods, adding a child to a household is associated with time demands that are not entirely accounted for by additional time spent in direct child care activities. This becomes accentuated as the children grow.

As with child care activities, the greatest impact comes with the birth of the first child, and the amount of household time spent in unpaid work goes down as the age of the youngest child goes up. This drop is slight when there is one pre-school child in the family, but more pronounced when there are two or more children. In one and

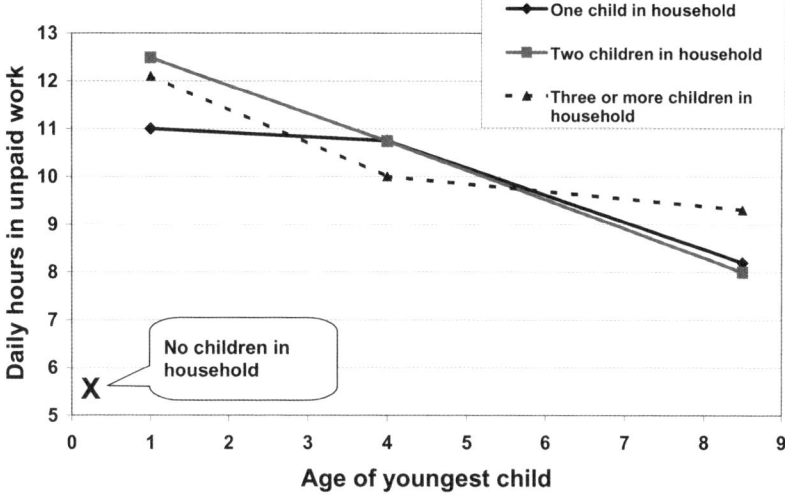

Figure 3.2   Predicted hours a day spent by couples jointly in unpaid work by number of children and age of the youngest child

Source: This graph shows fitted values from Table A2.

two child families, there is a big drop in unpaid work at school entry of the youngest child, again showing the enormous effect of school as an institutional substitute for parental time.

However, time in unpaid work does not follow the same pattern of steady decline as child care. When there is one child in the family, time spent in unpaid work time is almost the same whether the child is aged 0–2 or 3–4.

With one child aged 3–4, a household spends nearly five hours more in unpaid work than a child-free household. Three hours of that time (60 per cent) is in child care. This contrasts with families in which the youngest child is an infant, where 80 per cent of the additional unpaid work time is directed to child care alone. So the proportion of unpaid work associated with children that can be accounted for by time spent in direct child care drops as the child matures. It is apparent that the reduction in direct child care is counteracted by an increase in other types of unpaid labour. As the children age, the amount of household time spent in cleaning, washing, cooking or shopping grows as the need to provide direct child care falls. This shows that looking only at the amount of time spent in child care underestimates the impact of children on time demand.

As with direct child care, there are significant economies of scale with the amount of household unpaid work associated with additional children. When there is an infant in the family, household unpaid work in a family with two children is only 20 per cent more than the additional unpaid work time already associated with the first child. After this age, the marginal differences are so slight that the amount of unpaid work done in a family with a youngest child aged anywhere between 3 and 11 years is almost exactly the same whether there is child or two.

Families with three or more children again present something of a puzzle. Among three-plus child families with a youngest child aged 3–4 years, the drop in unpaid work time is even more pronounced than the drop in child care time reported above. Whereas families with three or more children do less child care than two-child families, but more child care than one-child families, families with three or more children do less total unpaid work than families with either one or two children. I discussed above how time in child care sub-categories contributed to this effect. Again limiting the sample to parents only, I find that families with one child over 2 undertake a higher proportion of housework to child care than do larger families. One-child families have a statistically significant greater input into domestic labour than larger families (except three-plus child families with a youngest child over school entry age). Three-plus child families with the youngest child not yet at school make time savings in all sub-categories of unpaid work. But the marginal differences are greater after school entry. When the youngest of three or more children is at primary school, the families do about an hour more unpaid work than one- or two-child families, a similar proportional increase as in direct child care time. More than half of this time is directed towards housework, and the rest is accounted for by incremental increases in physical care, interactive care and, most substantially, travel and communication.

However, while the total time cost of children does vary with the number and age of children in a family to some extent, it is the mere presence of children that has the most profound impact. Compared to households without resident children, *every* family configuration is associated with a substantially increased time allocation to unpaid work. This raises the question of how these households cope with such enormous time demands. I next look at how children impact on household time allocation to total work, both paid and unpaid.

Total Productive Activity

The term 'total productive activity' is used interchangeably with 'total paid and unpaid work', and (as explained in Chapter 2) is an aggregate measure that includes paid work, domestic labour and child care. It is intended to capture any increases in employment time, and in non-child care unpaid work, that are associated with the presence of children, in addition to direct child care. The running of most households typically entails a large input of paid and unpaid labour. The average total paid and unpaid work (as primary activities) for a couple family on the mean income, in which both adults are aged 35–44 and have no post-school qualifications, with no disabled family member and no children, on a weekday, is 17 hours a day (see Figure 3.3). The impact of the age and number of the children on this total paid and unpaid work time is markedly different than it is on its subsets, unpaid work and child care. First, the absolute impact is much less. The addition of a first child adds two hours to the daily work commitment of a couple. As a result, the proportional impact is also diminished. Following parenthood, there is an increase in total paid and unpaid workload of about 12 per cent of that already undertaken by childless couples. This is much less than the nearly 100 per cent increase in unpaid work time identified above.

Second, marginal time differences by number of children do not operate in quite the same way. Having two children, with one less than 2 years old is associated with

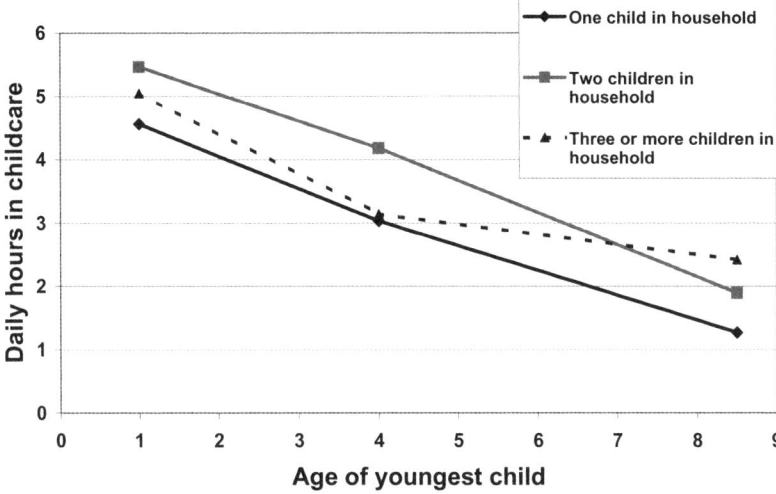

Figure 3.3   Predicted hours a day spent by couples jointly in total (paid and unpaid) work by number of children and age of the youngest child

Source: This graph shows fitted values from Table A2.

double the additional productive activity time of having only a single child under 2. There are no apparent economies of scale in this instance. Such a family averages over four hours more productive activity than a couple-only household. At a total of nearly 22 hours a day, this is the family type associated with the highest amount of total work. So in contrast to the findings for unpaid work, in which the birth of the first child occasioned the biggest impact, total household productive activity is almost equally affected by the presence of a second child (when the youngest is under 2).

In all other family configurations, total work loads do appear to show the effects of economies of scale, so that the extra time requirements associated with subsequent children are not as great as with the first. Echoing the findings for unpaid work, when there are three children in a family with a youngest child aged 0–2, the parents' combined total workload is less than when there are two children. Similarly, in families with a youngest child aged 3–4, there is very little difference in total work when there are one or two children in the family, and having three children is associated with less work than either. When the youngest child is of primary school age, having two or three or more children makes almost no difference at all to household time in total work.

A third point to note about household time in total productive activity is that one-child families show a time-use pattern distinct from larger families, which was not apparent for household time in unpaid work and its subset child care. In one-child families, there is a significant increase in productive activity when the child matures from infancy (0–2 years) to toddler-hood (3–4 years). When an only child enters school, total household productive activity declines to a level only one hour and 12 minutes more than a childless household. In contrast, two- and three-plus child families do about an hour less total productive activity when the youngest child grows from 0–2 to 3–4, and then level out at about two daily hours more than a child-free household. Once the youngest child is at school, total household productive activity in a one-child family is about an hour less than in a larger family. Unlike the results found in relation to unpaid work and child care, in families with two children, parents' total work does not drop much at school entry age. In families with three or more children it does not drop at all. So in total productive activity, unlike unpaid work, there does appear to be a significant time saving with one child compared with more.

When children are born to a household, the additional productive activity is less than the additional unpaid work associated with children. So rather than adding the child care and unpaid work demands associated with children onto the time already allocated to paid work, households redirect time resources to children. Time reallocation to the unpaid work associated with children from other activities is a major way in which households cope with the time demands of children. This is analogous to the redirection of monetary expenditure away from, for example, restaurant meals, following the addition of children to the household. In economic terms, the families are responding to the relatively higher price of going out by substituting other types of expenditure. There is, similarly, substitution towards the unpaid work associated with children from other types of time-use. I have already shown that as with money, time resources are subject to economies of scale. The analysis to this point has shown that these operate mainly within the time-use categories of direct child care and housework. Time resources are also subject to substitution effects. The major sources of redirected time are paid work, sleep and leisure (see Table 3.2).

Childless couples in the reference category supply an average of just over nine hours a day per household to the labour market. Couple households with child aged 0–2 spend between three and four hours a day less time in paid work. In two-parent households where the youngest child is of school age, labour supply is lower by about an hour and a half. The effect of number of children is not straightforward. It could be expected that the more children, the less household time would be devoted to paid work, and certainly, when the youngest child is 0–2 years, having three-plus children is associated with the greatest reduction in paid work time (nearly four hours a day). But in other age groups the number of children was not associated with an incremental reduction in paid work. It seems that although households with children substitute unpaid work for paid work, the age of the youngest child is crucial, so that there is no monotonic relationship with family size. No significant reduction in household labour supply was associated with the presence of one child aged 3–4, or with two children aged 5–11.

All households with children, except those with one child aged 3–4, spend at least an hour a day less in recreation than childless households. The number of children is at least as strong an influence on the reallocation of recreation time as is the children's age. In one-child families recreation time is reduced by more than an hour per day. In larger families the reduction is greater.

Personal care is comprised of grooming, eating, washing and sleeping. By far the largest component of this activity category is sleep. All couples with children spend

Table 3.2   Coefficients of hours a day spent by couples jointly in employment, personal care and recreation by number and age of children

	Employment		*Personal Care*		*Recreation*	
Number and Age of Children						
Constant term	9.09	***	22.24	***	7.46	***
Youngest 0-2						
1 child	-3.74	***	-0.87	**	-1.26	**
2 children	-3.03	***	-1.77	***	-2.55	***
3+ children	-3.97	***	-2.14	***	-1.39	***
Youngest 3-4	-1.99		-1.69	**	-1.37	
1 child						
2 children	-2.73	***	-1.53	***	-2.30	***
3+ children	-2.60	***	-1.97	***	-1.68	**
Youngest 5-11	-1.45	**			-1.01	*
1 child			-0.78	**		
2 children	0.33		-1.33	***	-1.36	***
3+ children	-1.46	*	-1.13	***	-1.39	***

* P-value<0.05 ** P-value<0.01 *** P-value<0.001

Source: Author's calculations of ABS TUS 1997. The figures in this table are drawn from Table A3.

less time in personal care than childless couples. Loss of adult personal care time continues well beyond the children's infancy, and the number of children affects adult time in personal care more strongly than does the children's maturity. In families with pre-school children, personal care time decreases steadily with each additional child. Unlike most other types of time allocation, there is a direct relationship between the number of children in a family and the reduction of adult sleep time.

Summary of Time Allocation by Couples Jointly

So far, this chapter has at looked the impact of children on couples time-use jointly. First, I identified the large amount of child care and domestic labour that a family must incorporate into its daily schedule following the birth of a child. Households with children have substantially more committed time to fit into each 24-hour period than childless households. This raises questions about how couples accommodate such demands. I found that they cope in several ways. They give up other types of work, most notably reducing their time in the paid work force. However, the reduction in paid work time is not traded off against unpaid work time on an hour-for-hour basis, so households with children also work longer in total than childless households. They not only substitute one type of work (unpaid) for another (paid), they also find time from other, non-work, sources. They take time that childless couples can spend in recreation and recuperation and allocate it to productive activities. This latter effect is durable. As children grow, the difference in *paid* work time between households with and without children lessens. Most households that have withdrawn time from paid employment following the birth of children reallocate much of it back to paid work as the children mature. This does not apply to recreation and recuperative time. The loss of parental time in rest and leisure in comparison with childless couples remains constant as the children grow, and continues well past the children's infancy.

I also investigated how couples manage the direct time demands of more than one child. I found that most often a similar amount of parental time is invested in child care despite variation in the number of children. With regard to daily workload, the difference between having children and being childless is far greater than the differences that result from variation in family size, and the age of the child is of more importance than the number. Broadly speaking, within youngest-child age bands a constant amount of parental time is shared by however many children are present in the family. There are some exceptions – one-child families manage a higher time allocation to housework than larger families, and two-child families spend more time in physical and interactive care than larger families. Child-related travel and communication time is less amenable to economies of scale and goes up steadily with each extra child.

So as with monetary expenditure on children, there are considerable apparent economies of scale in time cost. Research into the financial cost of children has found that households redirect spending towards child-related expenditure. Similarly, there is substitution towards the unpaid work associated with children from other types of time-use. The question now is how parental couples, with all the possibilities of sharing or specialisation open to them, distribute the extra work time requirement occasioned by the presence of children.

Time Allocation by Men and Women Individually

Child care

Figure 3.4 shows male and female time spent in child care separately. It makes clear that women contribute the overwhelming bulk of the large household time allocation to child care attendant on the birth of the first child. A first child adds over three and a half hours of child care to a woman's workload. The birth of a child adds just on an hour to the workload of a man.

Recall that the increase in household time in child care following the birth of a first child was four and a half hours. If this graph of the individual contribution of men and women is compared with household time in child care, it can be seen that the pattern for females closely echoes the household pattern. Male participation in child care is so subsidiary to female it hardly contributes to household allocation of time to direct child care activities. In families with a youngest child not yet at school, women are contributing over three-quarters of the household time allocation to child care activities. When the youngest child is aged 5–11, women are contributing about 85 per cent of the total household time allocated to child care. It is clear that specialisation by sex is profound.

Again echoing the household level analysis, the age of the youngest child has the greatest effect on the time women spend in child care. In one-child families, this

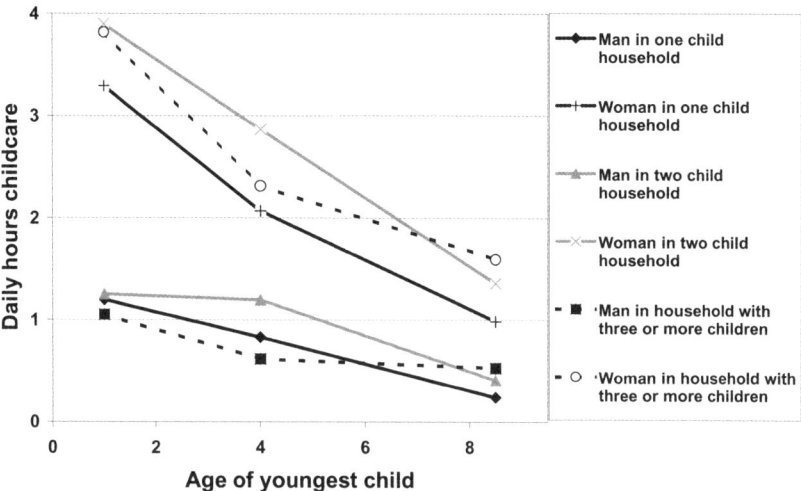

Figure 3.4   Predicted hours a day spent by men and women in child care by number of children and age of the youngest child

Source: This graph shows fitted values from Table A6.

time declines steadily and substantially as the child matures. When the child is 3–4, female time in child care is just over two hours a day. When the child is at primary school, female time in child care is just over an hour a day.

With the addition of subsequent children female child care time increases only marginally. Compared with a one-child family, when the youngest child is aged 0–2, the presence of either one or two older siblings is associated with an additional half-hour in female child care. Having several children increases a mother's daily child care time by only 12 per cent beyond that associated with a single child. Not only does absolute time in child care decrease as children mature, but also, economies of scale are less pronounced. They are still substantial, however. Each extra child in a family in which all the children are above school-entry age adds about 15 minutes to a mother's child care time. This amounts to a proportional increase of 25 per cent between a mother of one child and a mother of three-plus children. With a 3–4-year-old youngest, there is more female time spent in child care when there are two children than when there are three or more children. The marginal differences in child care time are even less for men. Indeed, when there is a pre-school child in the family men's participation in child care is lowest when they have three children. Men are most involved in child care when they have two children.

At the household level analysis, I found that the types of child care time most subject to apparent economies of scale are physical care and interactive care. Time-use data allow investigation into how these time resources are allocated by each partner in a couple. As with the analysis at household level, the sample for this part of the analysis was restricted to families with children, to see how these activities are divided between spouses.

Interactive care

Recall that households with three-plus children in which the youngest is not yet
at school spend less time in interactive child care than smaller households do.
Analysing the results separately by sex reveals that this decreased interactive time
largely reflects male behaviour (see Table 3.3). Both in families with an infant and in
families with a toddler, men spend the least time talking to, playing with, reading to,
teaching or reprimanding their children when they have three or more children. The
drop is proportionately quite high. When there is an infant in the family, fathers of
two spend nearly twice as much time interacting with their children than do fathers
of larger families. Men with one toddler spend about 20 minutes a day interacting
with them, while fathers of three or more share less than four minutes a day in these
activities amongst all their children. Maturity of the children is also associated with
less paternal interactive child care, with fathers in all families of school age children
averaging between six and just under 10 minutes a day in these activities.

Women's behaviour also contributes to the lowered household aggregate time in
interactive care in larger families, in that mothers of three-plus children who have
an infant spend less time playing and talking with children than do mothers of one
or two children. Mothers of two, the youngest of whom is 3–4 years, spend more
time in interactive activities than either mothers of one or mothers of three or more
children. However, the decrease in time is proportionately lower than for men. The
age of the children influences maternal time in interactive care more strongly than
the number of children does. For each youngest-child age range, a similar amount of
maternal talking and playing time is shared between all children present.

Table 3.3   **Coefficients of minutes a day spent by fathers and mothers in
              couple-headed households in child care sub-categories by number
              and age of children**

	Physical Care		Interactive Care		Travel		Passive Care	
	Father	Mother	Father	Mother	Father	Mother	Father	Mother
Constant term	12.76 ***	111.30 ***	25.02 ***	56.08 ***	0.57	8.00	22.54 ***	31.69 ***
Youngest 0-2								
2 children	7.95 *	22.26 **	4.875	3.88	3.00	10.42 **	-4.24	-5.27
3+ children	4.66	6.43	-8.64 *	-9.98 *	6.37 *	26.45 ***	-7.07	5.62
Youngest 3-4								
1 child	-8.79	-66.70 ***	-4.51	-0.14		16.03 **	-14.13 **	-16.92 *
2 children	4.43	-42.58 ***	-1.20	-14.61 **	6.75 *	25.98 ***	-8.76	5.27
3+ children	-8.43	-79.71 ***	-21.12 ***	-9.07	7.51 *	29.98 ***	-11.41 **	4.97
Youngest 5-11								
1 child	-20.63 ***	-108.03 ***	-16.79 ***	-20.41 ***	3.57	8.51	-20.50 ***	-22.16 ***
2 children	-17.26 ***	-98.25 ***	-18.58 ***	-22.68 ***	5.26 *	20.39 ***	-13.60 ***	-17.66 ***
3+ children	-13.40 ***	-87.93 ***	-15.47 **	-19.02 ***	6.76 *	25.11 ***	-17.17 ***	-17.92 ***

* P-value<0.05 ** P-value<0.01 *** P-value<0.001
Source: Author's calculation of ABS TUS 1997. The figures in this table are drawn from Table
A9.

Physical care

The lower physical care in three-plus child families found at the household level results mainly from female behaviour. Father's time in physical care is on average less than a quarter of mother's time in physical care, and apart from a slightly increased input in two-child families with an infant, in families that contain a pre-schooler male physical care time does not vary significantly with family size. In contrast, mothers of two spend considerably more time in physical care than mothers of larger families – 16 minutes more a day when the youngest is an infant, and 37 minutes more a day when the youngest is aged 3–4. The greater child care time found in two-child families with a pre-schooler can be attributed largely to female physical care.

Travel and communication

While mothers undertake the overwhelming bulk of travel and communication associated with children, this is the one child care activity to which fathers in larger families consistently spend more time than do fathers in smaller families. All fathers of three spend significantly longer driving or discussing their children than fathers in smaller families in the same youngest-child age band. In contrast to paternal interactive care, which seems to diminish as families increase in size, there is more male participation in travel and communication as families get larger. The amount of time is small, however, and at its highest (in three-plus families with a toddler) is just less than nine minutes a day.

This maximum male contribution to child-related travel and communication is nearly equivalent to the minimum female time allocation to the same activities. Mothers of one infant spend eight minutes a day in travel and communication associated with children. The average for all mothers is 26 minutes a day. Travel and communication is the only child care sub-category apparently immune to economies of scale. Echoing the results found at the household level, maternal time in these activities increases steadily with each child, and is influenced by number of children more than by their age. In families with a youngest aged 0–2, each extra child is associated with a more than 100 per cent increase in maternal travel and communication time. Again reflecting the findings at the household level, these are the maternal child care activities that are least likely to reduce as the children mature. For households, child-related travel and communication is the highest single child care time category when children are aged 5–11. The sex disaggregation shows this is also the case for women but not for men. The highest single child care allocation by fathers of school age families is nine and a half minutes a day spent in interactive care.

Passive care

Variation in passive child care time throws no light on where economies of scale are to be found. Much of the variation is not statistically significant, and those that are relate more to the child(ren)'s maturity than to family size. As at the household level, this largely results from the fact that passive child care is rarely performed as a primary activity. Passive supervision of children is most often carried out (as secondary activity) while the parent is also doing something else. This is explored in Chapter 4.

Unpaid work

The household-level analysis found unpaid labour increased beyond the amount accounted for by the new demand for child care. As the foregoing has made apparent, most of the direct child care time that a household must find following the birth of children falls to women. It is of interest to see whether this specialisation of activity on the basis of sex is also found in domestic labour other than child care.

For both sexes, unpaid work approximately doubles with the birth of the first child (see Figure 3.5). The proportional impact is slightly stronger for men. Having a first child takes male unpaid work time from a base of 50 minutes a day to two and a half hours a day. However, women in any family configuration do three to four times more unpaid work than men do. The constant term for time spent in unpaid work by a woman in a couple household in which the male works full time is four and a half hours a day. There is a predicted increase of nearly four hours upon first motherhood, of which, as we have seen, three and a half hours is direct child care. The presence of one child therefore brings total female unpaid work time to just less than eight hours a day. So although the proportional increase in unpaid work time attendant upon parenthood is slightly higher for men, the most striking thing is that the absolute quantity of extra hours is so much greater for women. As with direct

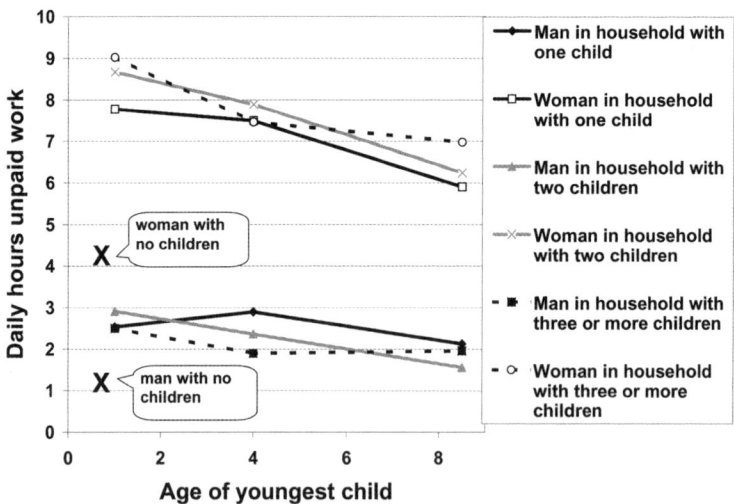

Figure 3.5   Predicted hours a day spent by men and women in couple-headed households in unpaid work by number of children and age of the youngest child

Source: This graph shows fitted values from Table A6.

child care, female input accounts for about three-quarters of the unpaid work time found at household level following the birth of a first child.

Further, the increase in male unpaid work time following the birth of a baby is allocated to direct child care and not to other forms of unpaid labour. This was the case in almost all family configurations. The exceptions were two-child families with a youngest aged 0–2, and families with an only child aged between 3 and 11 years. In these families there is a small but significant increase in male unpaid work beyond that allocated to child care. In all other families, the increase in male time in unpaid work was entirely directed into child care activities. In contrast, mothers of an infant allocate nearly half an hour more than childless women to unpaid work as well as three and a half hours to direct child care. Although female unpaid work declines as the children grow, the fall is not as steep as for child care alone, because for women the proportion of other domestic tasks to child care increases as the child matures.

In this instance, the female pattern for all unpaid work differs more from the household pattern than is the case with child care only. While women seem to benefit from powerful economies of scale, they do not show the drop in unpaid work that was associated with having three-plus children at the household level. For women, the impact of family size is at least as important as the age of the children. Each extra child is associated with slight increase in female unpaid work time at every age except when the youngest child is aged 3–4.

Broadly speaking, male time allocation to housework is higher the fewer children they have. When men are fathers of an infant, slightly more unpaid work is associated with having two children, but having three children is associated with the same amount of unpaid work as is one child. Men with a youngest child aged between 3 and 11 do more unpaid work when that child has no siblings than when there are more children in the family. These findings show that the male withdrawal of unpaid work in larger families is strong enough to outweigh, at household level, the increased input of resources by women with additional children. The increase in male time in unpaid work following the transition to parenthood is almost entirely composed of time spent in direct child care, so that it is women who contribute almost all the increased household time spent in shopping, cleaning or cooking. This implies that domestic inequity in the amount of housework performed is exacerbated by the presence of children.

Total productive activity

The increase in total female workload (total productive activity) following the birth of the first child is about an hour and a quarter a day. A new mother does about nine and a half hours a day total paid and unpaid work as a primary activity, 16 per cent more work than a woman without children. A second child brings maternal workload to a total of 10½ hours (about 20 per cent) more than the average workload of childless women. The major influence on women's total time in productive activity is whether they have a child at all, and what age the child is.

The presence of a first child does not predict quite as large an increase in men's work time as in women's. At 50 minutes, the time impact on men's total work of a

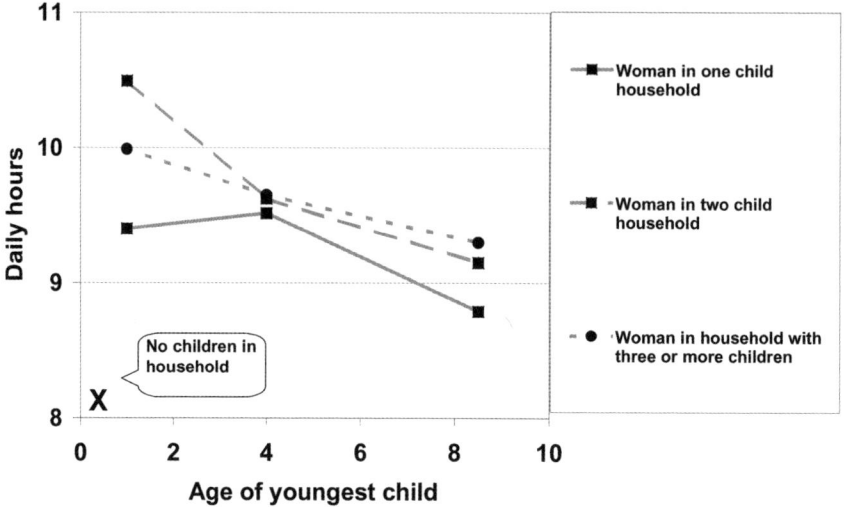

Figure 3.6   Predicted hours a day spent by women in couple families in total (paid and unpaid) work by number of children and age of the youngest child

Source: The figures in this table are drawn from Table A6.

first child is 75 per cent of the time impact on women. It also represents a somewhat smaller proportionate increase on male workload than for females. The birth of a first child is associated with an increase in male workload of 10 per cent (see Figure 3.6).

However among men, having two children, the youngest aged 0–2, is associated with a doubling of the impact of one child the same age. This brings the proportional increase over a childless man's workload to 20 per cent. Having an only child aged 3–4 years brings the same increase in total work time. Apart from the family configurations just mentioned, there is little variation in male workload with age of the youngest child or with more children in the family. With a youngest child aged 3–4, men spend least time in total work when there are three or more children, and most time when it is an only child. Parenthood brings greater work commitment for both sexes but, as at the household level, this is in total much less than the increase in unpaid work and its subset child care (see Figure 3.7).

Households reallocate time towards unpaid work from paid work, recreation and leisure, and personal care activities. The following analysis investigates how this reordering of time commitments is divided between couples.

Sources of Parental Time

Paid work

Overwhelmingly, it is women's employment time that is affected by family configuration. Household reduction in paid employment following the birth of

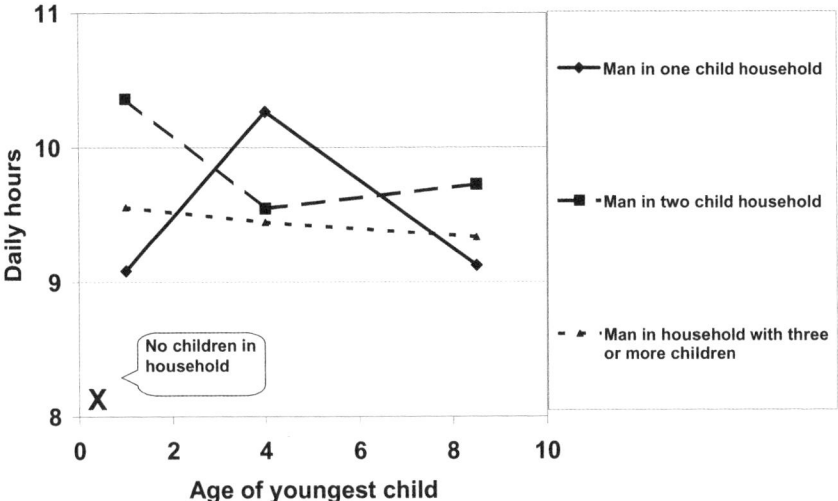

Figure 3.7 **Predicted hours a day spent by men in couple families in total (paid and unpaid) work by number of children and age of the youngest child**

Source: The figures in this table are drawn from Table A6.

children is caused by female behaviour. Male employment time is largely unaffected, either positively or negatively. In only two family configurations is male time spent in paid employment significantly different from that of childless men. Men with one infant spend an hour less in paid work than childless men, and men with two children at primary school spend about 50 minutes more. In contrast, *all* mothers spend statistically significantly less time in paid employment than childless women. Both the age and the number of children influence this time reduction. Predictably, the younger the child(ren), the less time a woman spends in paid employment, but also, having three or more children is, with an infant or primary school age youngest, associated with a further hours reduction in paid employment time than having fewer children in the same age group (see Table 3.4).

So both the age and the number of children strongly influence a woman's time in paid employment. Also, the more children a woman has, the longer the period of time she will have a youngest child in an age group associated with low work-force participation. Further, even when her youngest goes to school, the mother of a larger family reallocates significantly less time to paid work than mothers of smaller families. So following motherhood, women simultaneously increase their daily workload and forego access to income, and the more children a woman has the stronger is this effect.

Personal care

Typically, Australian women spend longer in personal care activities, including sleep, than Australian men do. Men with no children average nearly an hour less personal

Table 3.4   Coefficients of hours a day spent by men and women in couple families in employment, personal care and recreation by number and age of children

	Employment		Personal Care		Recreation	
	Men	*Women*	*Men*	*Women*	*Men*	*Women*
Number and Age of Children						
Constant term	5.83 ***	3.50 ***	10.89 ***	11.58 ***	3.53 ***	3.56 ***
Youngest 0-2						
1 child	-0.98 **	-3.07 ***	-0.45 **	-0.61 ***	-0.08	-0.90
2 children	0.31	-3.18 ***	-0.61 ***	-1.35 ***	-1.01 **	-1.42 ***
3+ children	-0.51	-4.07 ***	-0.75 ***	-1.74 ***	-0.35	-0.74 ***
Youngest 3-4						
1 child	0.16	-2.74 ***	-0.85 ***	-1.26 ***	-0.70	-0.39 ***
2 children	-0.35	-3.25 ***	-0.53 **	-1.12 ***	-0.74 **	-1.19 ***
3+ children	0.07	-2.85 ***	-0.75 ***	-1.47 ***	-0.46	-0.89 ***
Youngest 5-11						
1 child	-0.30	-1.61 ***	-0.45 **	-0.65 ***	-0.36	-0.39 *
2 children	0.86 **	-1.78 ***	-0.77 ***	-0.88 ***	-0.45 **	-0.63 ***
3+ children	0.15	-2.45 ***	-0.55 ***	-0.83 ***	-0.30	-0.76 ***

* P-value<0.05 ** P-value<0.01 *** P-value<0.001

Source: Author's calculation of ABS TUS 1997. The figures in this table are drawn from Table A7.

care than women with no children. But parenthood considerably narrows the gap, and the more children there are in the family, the stronger this finding is. The sleep loss for men ranges from 20 minutes to 50 minutes a day, but there is no pattern that relates to family configuration. In contrast, women lose between 36 minutes and one hour a day, and each child further erodes maternal personal care time.

Recreation and leisure

The effect on recreation time following parenthood is also very different for each sex. The only family configuration that is associated with a reduction in male leisure and recreation time is two children. The amount of leisure time lost becomes less as the children mature, but having two children of any age is associated with a reduction of male leisure time of between 27 minutes and an hour. As no other family configuration has this effect on male time, and a similar pattern of reduction was not found in personal care time or in employment time, it seems that this recreation and leisure time is the source of the greater paternal child care time identified above in families with two children.

When their children are preschoolers, women also sacrifice most leisure time when they have two children. But there is a major difference on the basis of sex: whereas men are losing leisure time only when they have two children, women are losing it in all family configurations except one-child families with a toddler. In all cases the lost female leisure and recreation time is greater than the male.

Summary of Results of Time Allocation by Men and Women Individually

The household time adjustment attendant upon parenthood is distributed very unequally by sex. The increase in workload (and the concomitant loss of recuperative and leisure time), and the substitution from one type of work activity (paid) to another (unpaid), are both more pronounced for women than for men. Therefore, the difference between being a mother and a childless woman is much more profound than the difference between being a father and a childless man.

Men with more than two children do less unpaid work then men in smaller families. This is the reason for the household finding that economies of scale in productive activity operate so effectively that three or more children in the family require less work time than two. Women increase unpaid labour time monotonically with each extra child, but in three-plus families the male withdrawal from these activities is so strong that at household level it masks the increased female input.

Having children, especially more than one, entrenches work specialisation by sex. In no family configuration is the gender division of labour equitable, but it is at its least extreme in childless households and in families with only children. In one-child households, fathers contribute child care and housework time beyond the male average. In two-child households, fathers contribute child care time beyond the paternal average. In larger households, fathers do neither. Also, the smaller the family, the more able are women to maintain time in the work force. Taken together, these findings suggest that (to personify a statistical average) if a woman wants to combine work and motherhood, and hopes for spousal contribution to the housework, having one child may be the most effective course. Two children would be the optimal number if she hoped for more paternal involvement in child care and were prepared to take on all the extra housework associated with children. More children than that and specialisation by women in unpaid labour and by men in paid labour is entrenched.

Time Allocation of Sole Parents

The analysis thus far involves a comparison among couple households, that is, childless couples and couples with varying numbers of children and ages of youngest child. Households that contained adults in addition to the couple were excluded from the sample. Finding a point of comparison for sole parents is less straightforward. Should sole parent households be compared to childless single person households or to couple households with children? To address that question, I ran further regression analyses. The first compared single parents with childless single people and with couple parents on the major dependent variables. The second took parents (single and partnered), and compared their time in child care. The results are shown in Tables A12 and A13.

Sole parents, the vast majority of whom are women, are of particular interest because, in their households, the functions of earning money and caring for children fall to one individual. Mothers in two-parent families shoulder more domestic responsibility than fathers, but in families with one parent, the demands of home and family are not even possibly a matter of intra-household allocation and negotiation, but simultaneous requirements of one person. The dilemma of whether to work for money or to undertake household production is a problem for all women, but it is particularly difficult for sole mothers (Orloff, 1996; Duncan and Edwards, 1997; Land and Lewis, 1997; Scott et al., 1999; Lewis, 2001; Ellwood and Jencks, 2002; McLanahan, 2002; Moynihan et al., 2002).

Compared to two-parent households, lone parents not only have reduced money, but they also have half the adult time resources available (Douthitt, 1992). However, while there is a great deal of research into the financial consequences of sole motherhood (for Australian examples see Bradbury and Jantti, 1999; ACOSS, 2000; Harding et al., 2001; Healey, 2002) information about the non-monetary side of the equation has been slow to emerge. With the exception of Douthitt (1992), it is only recently (Craig, 2005; Kalenkoski et al., 2006; Craig et al., 2006) that there has been systematic investigation into the extent of sole mothers' caring responsibilities and the implications for paid and unpaid workload. Tables A12 and A13 show the results of the analysis conducted for this book. The figures discussed below are fitted values calculated from these analyses.

Total productive activity

Single men and women do the same amount of (paid and unpaid) work in total. Being partnered but with no children is associated with an increase in workload for both men and women, with the effect being slightly more pronounced for women. Women who have male partners do more work in total than single women, whether or not there are children in the household. A partnered mother of two children, the youngest of which is under five, spends nearly three hours a day more time working in total productive activity than does a single childless woman. A sole mother of two, the youngest of which is under five averages two more hours a day total work than a single woman, nearly half an hour less than a partnered mother. In households with children aged 5–11 years, partnered parents both spend about an hour more time working than a single woman. In contrast, the total workload of a sole mother of a school-aged youngest is not significantly different from that of a single woman. Just being in a couple makes more total work for all. Whatever their motherhood status, partnered women do more total work than non-partnered women. It is not sole mothers who have the highest total workloads, but women who combine motherhood with partnership.

Unpaid work

The effect of partnership upon the unpaid work time of men and of women is diametrically opposite. Women's unpaid work goes up; men's unpaid work goes down. While single women do somewhat more housework and shopping than single men (22 minutes a day), being part of a couple increases female domestic work time by an hour a day, and decreases male domestic work time by over half an hour a day. So the composition of the partnership workload penalty is different by sex. For men it is comprised entirely of paid work, but the partnership workload penalty for women includes additional domestic labour.

Of interest here is whether this partnership penalty persists when there are children in the household. It would seem logical to predict that because sole mothers have no partner to assist with household work and child care, they do more unpaid labour than partnered mothers. However, sole mothers do no more unpaid work than do partnered mothers. Mothers of children of any age between 0 and 11 do an almost identical amount of unpaid labour whether they have a partner or not.

Indeed, the results suggest that if the housework component of unpaid work is separated from child care, men do not increase their contribution towards housework, shopping and other non-child care domestic tasks when there are children. Fathers in families with youngest children under 5 do no more housework than men in childless couples, and fathers in families with a youngest over 5 years old actually do *less* than men in childless couples. The absence of a male partner does not mean that sole mothers do more domestic work to make up for the lack of male input. On average, male contribution to domestic labour is a net deficit. Whether or not women have children, they spend longer doing housework and shopping when there is a man present in the household than when there is not. Contrary to the popular expectation that being a single mother entails more housework than being a married mother, the absence of a resident man slightly reduces the domestic burden upon mothers.

However, the other major component of unpaid work is child care, and it is this aspect of the responsibilities of sole mothers that cause the most social concern. I now turn to the question of how partnered parents and sole mothers allocate time to their children.

Child care

There is no statistical difference in the amount of time mothers are engaged in direct care of their children according to whether or not they have a partner. Fitted values of the time parents in two-child families with children under 5 commit to child care according to family structure suggest that both couple and sole mothers of two children under 5 average three hours a day in child care when it is counted as a main activity. Men in partnered families spend just over 50 minutes a day in primary child care. These results may seem to indicate that while sole mothers match the child care inputs of mothers with partners, children in couple families receive more care in total when the father's input is included. However, the nature of time with children is more complicated than this implies. This will be addressed in more detail in the next chapter.

Discussion and Conclusion

In this chapter, I loosely adopted a marginal costs approach to establishing the time costs of children, using a difference model of daily adult time in total paid and unpaid work as a basis for comparison of welfare across household types. I attempted this marginal cost comparison quantified in the metric of daily workload because the current social problem of balancing work and family is perhaps more about significant time constraints than about the scarcity of money resources. Expenditure equivalence scales aim to account for economies of scale in household size, price-like substitution effects and intra-household allocation of resources. I found that these were also of relevance to the daily time cost of children.

There are big entry costs to parenthood. There is a significant difference in time spent in productive activity between those who have children and those who do not. The biggest time demand comes with the birth of the first child. It is the decision to have any children at all that creates the largest time-commitment division. Daily workload varies more between non-parents and parents than between parents of differing family size. Parents do more total work than non-parents, and reallocate time from other activities to perform this extra work.

Incremental increases with each child are much less than between one and none. As with monetary expenditure on children, there are considerable apparent economies of scale in time cost per child. The amount of time allocated to child care with the birth of the first child is much greater than the amount allocated for each additional child. There is a fall as the children age, particularly at school entry.

Monetary expenditure on children goes beyond the amount needed to buy child-specific goods and services. They occasion extra outlay on shared items such as housing and transport. Similarly, the extra time requirements of children go beyond straight child care. Their presence is also associated with increased time in other unpaid labour. As children age, proportionately more time is directed to the associated unpaid work than into actual child care. Research into the financial costs of children has found that households redirect spending towards child-related expenditure. Analogously, I found reallocation from sleep, personal care, paid work and recreation towards the unpaid work associated with children.

Household level analysis showed that the presence of children places an enormous demand on household non-market labour time resources. Of interest is how households distribute the extra work time requirement occasioned by the presence of children. The presence of children deepens the division of domestic labour. Women are contributing the bulk of the large household time allocation to child care and unpaid work that follows the birth of the first child. Women forfeit sleep, leisure and time in the labour market. Fathers' total work load is lower than mothers'. Compared to childless men, fathers allocate extra time to paid work and some child care. Few are contributing more domestic labour. This calculation of the difference in time commitment between families with no children and families with different numbers of children gives a measure of the heavy time pressure associated with work and family demands. It shows that the time increases associated with children found at household level are largely met by adjustments on the part of women. Specialisation by sex is the major response to the additional household time demands of parenthood. In the next chapter, I investigate whether there is also a difference in the total work burden upon men and women, and the type of child care performed by each sex.

Chapter 4

Gender Equivalence or Hidden Inequity?[1]

In this chapter I tease out some of the more subtle aspects of how children and their impact upon workload differ for men and for women. I find that there are dimensions of double activity, responsibility for children, child care task allocation and total workload that previous analyses have overlooked, which show that the time effects of parenthood differ profoundly by sex. There are differences between mothers and fathers both in the total work burden they shoulder, and in the way they spend time with children in relative terms.

Background

There are two divergent views on how modern households currently manage the requirements of paid and unpaid labour. One is that the historical trend over time is to a gender convergence. Proponents of this approach argue that there is decreasing specialisation in activity on the basis of sex. They also contend that currently there is broad equality between men and women in the time spent in total productive activity and leisure (Robinson and Godbey, 1997; Gershuny, 2000; Beaujot, 2001; Gershuny et al., 2005; Bianchi et al., 2006). The opposing camp argues that while women are becoming less specialised in the way they allocate time, men continue to specialise only in paid market activity. Although large numbers of women are participating in paid work, they are also retaining responsibility for the unpaid domestic work. Therefore, it is argued, wives are therefore working much more than their husbands, to meet the demands of this dual burden (Hochschild and Machung, 1989; Schor, 1991). The revolution has stalled.

There has been a vigorous debate over which version is the more accurate, which is as yet unresolved. Each side asserts that the other sides' perception of the gender division of labour is distorted, the first holding that the disparity between male and female workloads is widely overestimated, and the second that it is played down. This chapter uses the detailed time–use measures set out in Chapter 2 to shed new light on this issue.

---

1   Parts of the argument in this chapter appear in Craig, Lyn (2006): 'Does Father Care Mean Fathers Share? A comparison of how mothers and fathers in intact families spend time with children' *Gender and Society*, 20 (2): 259–281 © 2006 Sociologists for Women in Society.

Thesis: Gender Convergence in Productive Activity

Many argue that men and women are becoming less specialised in the way they allocate time and that daily activity distribution is becoming more androgynous (Gershuny and Brice, 1994; Robinson and Godbey, 1997; Gershuny et al., 2005; Bianchi et al., 2006). Implicit is a view of marriage as a partnership of two equals. 'The symmetrical family' is one in which marital partners have equal status and responsibility (Young and Wilmott, 1973). The ideal marriage is increasingly seen as an equitable arrangement in which husband and wife both contribute financially and emotionally (Giddens, 1991; Beck and Beck-Gernsheim, 2002).

Liberal feminists, who see women's equality as resulting from improved opportunities in education and employment, have historically supported this model. 'Generally, feminists look to a future in which women and men are equal in opportunity, in respect and in the burdens they carry' (Bergmann, 1986). They argued that the best way of achieving equity between the sexes is by improving the economic power of women. On this view, the priority for feminist action lies in access to employment, and discrimination and unfair practices in the workplace are the primary targets for reform. The solution to domestic disadvantage is greater economic independence. When women earn more, they will have more power in the family, so equality between husband and wife in domestic labour should be easier to achieve (Bergmann, 1986). Liberal feminists see women's access to economic resources on the same basis as men as the key to wiping out female disadvantage (Hartmann, 1981).

So it was predicted that with the entry of large numbers of women into the work force there would be a complementary increase in male involvement in domestic activities. Oakley points out that assumptions of growing sex equity partly arise from simple logic. One consequence of women's increasing time in paid employment time *should* be men's increasing domesticity (Oakley, 1985). It was assumed that the family would act as a self-balancing system in which, if one member withdrew from certain types of labour, another would take over that labour. This has been called the 'adaptive partnership model' (Meissner et al., 1975). In an early advocacy of this approach, Myrdal and Klein wrote in 1956 of how both sexes could have dual roles in paid and unpaid work. They presented data on the long hours spent by women on housework, and optimistically advocated an equitable six-hour paid work day for both sexes, arguing that 'both men and women would [then] have more time and energy to look after their home, an activity which can be a creative combination of work and leisure if one has neither too much nor too little time for it' (Myrdal and Klein, 1968, p.192).

Research has found some evidence for the view that the division of domestic labour is becoming less pronounced. Some time-use analysts use a combination of description and prediction to identify signs of increasing androgyny and to argue that this means that over time there will be greater convergence in male and female activity. They note modest male adjustments in time allocation towards household tasks, and greater equality of total time investment by men and women in domestic labour, and note movement towards what may become a collaborative, unspecialised, model of household labour (Myrdal and Klein, 1968; Gershuny et al., 1994;

Robinson and Godbey, 1997; Gershuny, 2000; Beaujot, 2001; Gershuny et al., 2005; Bianchi et al., 2006). For example, using US time-use data to compare male and female activity patterns from 1965 to 1985, Robinson and Godbey identify a change towards gender homogeneity in productive activities, saying that women are doing more paid work and less unpaid work. They find that female time-use patterns are more like male and male time-use patterns are more like female. They argue that gender is becoming a 'false variable' that will progressively become less predictive of numerous characteristics including time-use (Robinson and Godbey, 1997).

The gender convergence viewpoint is also being applied to parental time with children. Like marriage, parenthood is not a stable ideal over time (Burgess, 1997; Pleck and Pleck, 1997). In the 1950s new ideas about mother-child bonding and attachment emerged, which created concern that women who worked outside the home were depriving their children of necessary care (Myrdal and Klein, 1968; Mitchell, 1971; Oakley, 1974; Rich, 1977; Oakley, 1985). In the ensuing decades, ideas about the optimal maternal input to children have remained contested (this will be discussed in more detail in Chapter 5). Similarly, ideals of fatherhood have altered. In western societies, the approved model used to be a sometimes-distant breadwinner; now it is a dad who involves himself in the direct care of his own children (Griswold, 1993). We are moving towards a social ideal of father as co-parent (Burgess, 1997; Pleck and Pleck, 1997; Cabrera and Tamis-LeMonda, 1999; Bianchi and Casper, 2000; Casper and Bianchi, 2002; Gerson, 2002; Castleman and Reed, 2003; Coleman and Ganong, 2004).

As well as description and prediction, there is a strong element of advocacy in the expert commentary on gender convergence in parenting practices. This comes from several different perspectives. Children who have involved fathers have better developmental outcomes than children whose fathers are less engaged (Lamb, 1997; Silverstein and Auerbach, 1999; Cabrera et al., 2000; Yeung et al., 2000). Others assert that greater involvement with children will bring rewards to the fathers themselves (see for example Petre, 1998; McMahon, 1999). Studies find evidence that men want to spend more time with their children (Schor, 1991; Russell, 1999; Milkie et al., 2004).

Also argued is that society will benefit from greater paternal involvement with children. This is from both a conservative position which advocates a return to the 'traditional' nuclear family structure with the male role as father valued and secure but mediated through the mother, and from a liberal perspective which sees social value in creating father-child bonds independent of mother-father bonds (Silverstein and Auerbach, 1999). Psychologists suggest that fathers who take a secondary role in parenting may develop relationships with their children that are reliant on the mother being present. This not only means children in intact families feel 'father hunger', but also, fathers who are active carers in intact families are better prepared to parent independently after divorce or separation (Furstenberg and Cherlin, 1991; Seltzer and Brandreth, 1994; Silverstein and Auerbach, 1999; Coltrane and Adams, 2003).

Increased father involvement is also seen as a way of dealing more equitably with the household demands on women (Hochschild and Machung, 1989; Folbre, 1994b; Joshi, 1998; Hewlett et al., 2002). Liberal feminists see female withdrawal from child care as a mechanism for encouraging male participation (Burgess, 1997) and many argue against the idea that one way of dealing with female marginalisation would be to give a higher social value to looking after children (see Hobson et al., 2002). For example, Bergmann contends that even if care were more highly valued, the tasks, and the associated economic vulnerability, would continue to fall upon women. Separating child care from the home and encouraging both partners to seek economic independence is a necessary interim step to achieve the long-term goal to establish a new ethic; that of sharing family care work between men and women (Bergmann, 1986).

So the idea that fathers should share in the care of their own children meets with a chorus of approval. But is it actually happening? Researchers have begun to look for indications that men are becoming more involved with children. Efforts to describe the extent to which co-parenting is actually being achieved come up against the unresolved question of what father involvement is, and how it should be measured. Ways of assessing fathers' involvement range from calculating the absolute quantity of time spent, comparing fathers' involvement to mothers', to investigating the kind and quality of care given. Pleck argues that 'positive paternal involvement' should be assessed using both qualitative and quantitative measurements (Pleck and Pleck, 1997). A tripartite model of paternal involvement that covers engagement, availability and responsibility has informed both qualitative and quantitative studies (Lamb et al., 1985; Lamb et al., 1987; Lamb, 1997; Cabrera and Tamis-LeMonda, 1999; Yeung et al., 2001).

There is also interest in what fathers *would* do if they had more time to spend with children. The 'availability hypothesis' holds that the more time a husband has to care for his children the more likely he is to do so (Cabrera and Tamis-LeMonda, 1999). This view implies that the current low levels of paternal care result from men's high level of engagement in the work force; if they were free to, they would spend as much time with children as mothers do. A related question is what fathers currently do in the time they *do* have available to care for children, which translates to a research question as to whether mothers and fathers spend their relative time with children in the same way. The 'availability hypothesis' was explored in a study that undertook a weekend/weekday comparison of paternal time with children. The researchers contend that a 'new father' role is beginning to emerge on weekends, and predict that the trend to more father care will continue (Yeung et al., 2001).

Antithesis: Gender Inequity and the Dual Burden

The identification of androgyny in productive activity – housework, child care and employment – is rejected by the opposing view on household allocation of paid and unpaid work. This view holds that sex remains central to the division of productive activity. Women are becoming less specialised in the way they allocate time, but men continue to specialise only in market activity. Women are still overwhelmingly responsible for housework and child care. The only change is that many have added employment to their responsibilities.

In a best-selling book that widely disseminated this perspective, Hochschild (1989) investigated how 50 households divided the responsibilities of paid and unpaid work. Identifying a widespread lack of domestic contribution by men, she suggested that women are largely responsible for household work, child care *and* the requirements of their paid job. The movement by women towards task androgyny met no corresponding movement from men. She characterised the take-up by women of paid work in the face of no adoption of domestic work by men as a 'stalled revolution' and popularised the term 'second shift', to encapsulate her contention that women are in the enormously stressful position of performing two jobs.

Quantitative studies of the division of domestic labour confirm that women perform a disproportionate amount of housework and child care (Meissner et al., 1975; Pahl, 1984; Hochschild and Machung, 1989; Bittman, 1992; Shelton, 1992; Baxter, 1993; Bittman and Matheson, 1996; Lamb, 1997; Harrington, 1998; Baxter, 2002; Bianchi, 2004; Bianchi et al., 2006). Researchers have found that the greater domestic equity identified by convergence theorists is much more due to women spending less time in household work than it is to men spending more (Bergmann, 1986; Buxter, 1993; Buxter, 2002; Baxter et al., 2005). Nor does this appear to be changing over time, as widely predicted. A recent Australian study found that a reduced gender gap in housework is still resulting from women dropping their time investment, not from men increasing theirs (Baxter, 2002). Women's time allocation remains less specialised than men's. The study also found no indication that the gender gap in *child care* is diminishing (Baxter, 2002). So while women are reallocating their time away from housework, and thus contributing to overall gender equality, they do not seem to be similarly pursuing the option of reducing time in child care (Bianchi et al., 2006).

Research has also established that there are persistent differences between men and women in the type of household tasks performed, and the conditions under which they are carried out. Men are more likely to help with domestic tasks than to be entirely responsible for them, and to participate in only a limited range of activities, such as gardening or car care. The household work they do undertake is likely to be more irregular and discretionary than that done by women (Oakley, 1974; Meissner et al., 1975; Bittman and Matheson, 1996; Sullivan, 1997; Baxter, 2002).

The picture of the overstressed and overcommitted working mother resonates with another view, that contemporary American society is experiencing a 'time famine' (Schor, 1991). There are only so many hours in a day. It is suggested that in western capitalist countries, in which time is viewed as a commodity, time feels like an increasingly scarce resource (Cross, 1993; Gershuny, 1999). There is a speed-up in the pace of life and there never seems to be enough time in the day (Rybczynski, 1991; Cross, 1993). Robinson and Godbey have identified responses to time 'famine', which they categorise as 'time-deepening' behaviour (Robinson and Godbey, 1997). More money can be found through working longer or more efficiently, but people cannot make more time. On the other hand, time can be stretched in ways that money cannot. Time-deepening behaviours include doing things faster, doing several things at once, scheduling activities more tightly, and cutting corners. Time-deepening behaviours allow people to crowd more activities into the same time frame (Robinson and Godbey, 1997).

Perception and reality

However, diagnosis of time famine is contested. Robinson and Godbey have conducted quantitative time-use analysis that suggests that while people adopt time-deepening behaviours and *feel* rushed, they actually do not have less leisure time over all (Robinson and Godbey, 1997). Therefore, they argue, time famine reflects changes in perception rather than actuality. The leisure people get under these circumstances may be of reduced recuperative value, but comparisons over time show that time in productive activity has declined, not increased (Gergen, 1991; Gershuny, 2000). Theorists of time allocation suggest that people are victims of their own aspirations to greater leisure and higher consumption (Becker, 1965; Linder, 1970; Rifkin, 1987). The feeling of harriedness is not real evidence of actual change towards more working hours (Robinson and Godbey, 1997; Gershuny, 1999; Gershuny, 2005).

Robinson and Godbey also use time-use analysis to argue that, currently, time spent by men and women in the US in work and leisure is broadly equal (Robinson and Godbey, 1997). This finding is supported by researchers in Britain who show that since the 1960s there has been a gender convergence in the amount of time dedicated to total work and leisure (Gershuny, 1999; Gershuny, 2000). Comparisons of time-use in other OECD countries replicate this finding (Beaujot, 2001; Bittman, 2004; Bianchi et al., 2006). On such evidence, the dual burden cannot be interpreted literally. Its truth may lie in the findings, discussed above, that women retain responsibility for domestic labour, but these studies challenge the assumption that they spend double the amount of time men do working. Robinson and Godbey contend that a broadly equal time investment in productive activity by men and women pertains cross nationally, except in Bulgaria, Italy and Austria, arguing that an 'invisible hand' keeps total productive activity more or less equal between the sexes (Robinson and Godbey, 1997). On this analysis, the claim that women are shouldering a double burden reflects perception rather than reality.

Conversely, many who hold the view that women have become less specialised in the type of productive activity they undertake, but that men have not, identify gender convergence as a mistaken perception. They argue that apparent gender convergence is a chimera and that there has been much more change in attitudes than behaviour (Hochschild and Machung, 1989; Bittman and Pixley, 1997; Dempsey, 1997; McMahon, 1999; Beaujot, 2001). While the ideal of egalitarian marriage has been widely accepted as desirable, and has come to inform sociological theories of modern relationships (Giddens, 1991; Beck and Beck-Gernsheim, 2002; Beck-Gernsheim, 2002), the actual sharing of household labour is not becoming more equal. Further, it is suggested that if ideology runs ahead of practice, the result may be that people feel like personal failures if their experience does not match the ideal (Joshi, 1998). There is evidence that marital equity is now such a strongly held social value that people define their domestic arrangements as fair even when objective analysis shows they are not (Baxter, 2000). A small amount of male assistance can be defined as sharing domestic work. In some cases, the willingness to participate if asked is counted as sharing (Thompson, 1991). It seems that many women add the psychological challenge of believing they are sharing the load when they are not to the physical burden of actually doing it all (Hochschild and Machung, 1989; Bittman and Pixley, 1997; Dempsey, 1997).

Summary

I have described two different positions on how couples are dividing paid and unpaid labour between themselves. The first view is that, currently, men and women are doing about the same amount of total productive activity and enjoy about the same amount of leisure time. Its associated contention is that there is a trend towards gender convergence in task allocation (Robinson and Godbey, 1997; Gershuny, 2000; Beaujot, 2001; Bianchi et al., 2006). The second view is that women are becoming less specialised in their productive activities, but men are not. Its supplementary claim is that women are therefore working long days (a 'second shift') to meet the accumulation of demands upon them (Hochschild and Machung, 1989; Schor, 1991). In addition, each side argues that people's perceptions on the magnitude and scope of the problem are distorted, the first holding that the disparity between male and female workloads is exaggerated, and the second that it is underestimated. Most previous analyses of time-use have not been comprehensive enough to investigate each claim fully.

Research Focus and Measures

In this chapter, I use detailed data from the TUS to shed greater light on the question of whether women are doing a 'second shift' while men are not, and how mothers' and fathers' parenting roles compare in relative terms. While the cross-sectional nature of the data does not allow analysis of historical trends, the measures I examine show dimensions of child care time that previous research has not examined, to answer the question of whether there is gender parity in total workload, and/or similarity in relative child care practice.

Magnitude of Time Allocation

Secondary activity (a)

One characteristic of the time cost of children is that a very high proportion of child care is done at the same time as other activities. About twice as much child care is done as a secondary, or simultaneous, activity than as a primary or main activity. Some time-diaries allow people to report simultaneous activities. However, respondents to time-diaries who are in charge of children, and who undertake another activity such as shopping, much more frequently record the shopping as their main activity than the child care they are also performing. This means that in most previous studies of unpaid work, child care has been significantly underestimated because it has only been counted when it was done as a main activity (Ironmonger, 1996). With few exceptions (Zick and Bryant, 1996; Craig, 2002; Ironmonger, 2004) earlier studies have excluded secondary activity from the analysis of time-use in connection with children. In this chapter I research total work time in a way that includes secondary activity (as described in Chapter 2).

Including secondary activity in the estimation of child care time is important for several reasons. First, it gives a fuller count of the amount of time parents commit to children. It allows calculation both of time that parents are engaged in child care and of the time that they are available to be called upon. This is important when considering the effect of children on adult time, as it shows how child care acts as a constraint on parents. While the presence of children may not require activity or direct intervention, it does prevent the carer from being elsewhere, and allows only certain types of other tasks to be undertaken. So including secondary activity in estimations of child care time reveals how being responsible for children acts upon parents as a constraint. Child care as a secondary or accompanying activity requires the parent's presence and at least part of their attention. It is time during which they cannot undertake activities where children cannot be present unless they arrange to substitute someone else's care for their own. So calculating secondary activity as well as primary is essential to full comparisons of how parents' lives differ from the childless, and whether the magnitude of the time impact of parenthood is different for men and women.

Experience of Providing Child Care

Secondary activity (b)

Counting secondary activity also allows some inferences to be made about the quality of the experience of providing child care, which allows a more finely graded comparison of the differential impact of parenthood upon men's and women's time. First, counting only the main task conceals how many activities are being done at once. Performing more than one work task at a time is often necessary because some jobs, such as cooking dinner and comforting a crying child, cannot be rescheduled (McMahon, 1999). Primary activity leaves such urgency and multi-tasking unrecorded. Including secondary activity in the measure of parental time acknowledges the density of activity associated with children. Second, recording simultaneous activity captures some of the subjective experience of responsibility associated with care of children (Sullivan, 1997). Even leisure time spent in the company of children, although it may be pleasurable, requires vigilance and attention. A picnic, for example, in the company of a child is very different from one without the responsibility of supervision. Finally, child care activities that could in themselves be pleasurable, like talking, bathing, playing, or reading aloud, can become less so if one's attention is simultaneously being claimed by other responsibilities. Attention to these aspects of secondary activity is a basis for comparing how the experience, not only the magnitude, of child care can differ by sex.

Relative task allocation

I take this gender comparison of the nature and experience of child care further. As discussed above, using TUS coding, child care can be divided into physical and emotional care of children, talking to/playing with/reading to/reprimanding children (interactive care), passive care of children, and travel and communication associated with children. Some of these tasks, such as feeding, dressing or transporting children have to be done at certain times, while others, such as playing or reading can be performed at the parents' discretion. Having disproportionate responsibility for those tasks that have to be done on schedule is an indicator of constraint. Additionally, some of the child care tasks are more pleasant than others.

Research has established that there are persistent differences between men and women in the type of household tasks performed. The household work men undertake is likely to be more irregular and time-flexible than that done by women. Male domestic tasks are disproportionately those such as lawn mowing which can be done at the man's discretion, whereas women's are typically those such as cooking, which must be done at a particular time (Baxter, 1993; Bittman and Matheson, 1996; Dempsey, 1997; McMahon, 1999; Baxter et al., 2005). Activities that have to be done to schedule are more constraining than those that can be fitted in around other activities.

In this chapter, I investigate whether similar differences in task allocation pertain to child care, by comparing time spent in each of the sub-categories of child care by sex.

Responsibility for the job of child care

Relatedly, there is a difference between having full responsibility for a job and giving occasional help. The role of helper is far less demanding. Research suggests that in many cases male help with domestic labour is not obligatory and routine, but a matter of choice (McMahon, 1999). Men may help with tasks, but the job remains the woman's responsibility. Even when both partners participate in an activity such as laundry, men are more likely to assist than to manage the whole job. If a woman cannot elicit assistance, she must do it herself (Dempsey, 1997). Further, women typically are assigned the role of manager of domestic responsibilities. Even in households that share housework it is the woman who must assume the responsibility for planning and organisation (Deutsch, 2000), which many describe as the most onerous aspect of domestic labour (Coltrane, 2000).

I am interested in whether this also applies to parental time with children. To look at this, I adapt a measure developed by Oriel Sullivan. She used time-use data to investigate whether respondents are more likely to be helping rather than taking responsibility for domestic labour, by calculating the proportion of time devoted to a particular task when alone. She argued that the more relative time in which the task is undertaken in the presence of others also doing it, the more participation in it could be regarded as auxiliary (Sullivan, 1997). I have applied the same principle to investigating whether child care is similar in this respect to domestic labour. The TUS provides information on where activities take place and who is present at the time. As discussed in Chapter 2, I used this 'company' information to calculate new variables showing who was present at the time child care was performed.

The aim is to see if there are gender differences in the proportion of child care time spent in sole charge of the children. There are three main implications if fathers' time with children is disproportionately spent in the presence of the mother. First, it means that the father is not functioning as a substitute for the mother's time. She is not able to use this time for other pursuits, including paid work. Second, it means that the mother is still taking the major responsibility for the job of child care, and the father is helping with it. It is therefore a qualitatively different experience. Third, the father-child relationship may be weaker if the mother always mediates it or acts as gatekeeper, and this may have detrimental consequences to independent father-child contact following divorce or separation (Furstenberg and Cherlin, 1991; Seltzer and Brandreth, 1994).

Quality of leisure

Parenthood is accompanied by an overall reduction in leisure time, as we saw in Chapter 3. What is unknown is whether mothers and father experience this reduction in the same way. The conclusion that if both paid and unpaid work is totalled, men and women on average enjoy a similar total amount of leisure, that some time use researchers have reached (Robinson and Godbey, 1997; Gershuny, 2000) pays no attention to the quality of that leisure. Other time use researchers argue that an adequate picture of gender equality in leisure cannot be obtained from merely adding up the amount of leisure enjoyed as a primary activity by each sex (Mattingly and Bianchi, 2003; Bittman and Wajcman, 2004). Investigating whether leisure is child free, and how long episodes of leisure last gives a more meaningful picture. On these measures there *is* a gender gap in leisure: women enjoy less leisure without children present (adult leisure) than men, and their leisure time is more often interrupted than men's leisure time (Mattingly and Bianchi, 2003; Bittman and Wajcman, 2004). In this chapter I am interested specifically in the extent to which *child care* responsibilities encroach upon the leisure of fathers and of mothers. I present a gender comparison of the proportion of total leisure time during which parents are not performing child care as a secondary activity.

In the final section of this chapter, I look at how sole parents' time on the measures above compares with that of partnered parents.

Results

In Figure 4.1 the solid lines represent total productive *primary* activity and the dotted lines show productive activity *as either a primary or secondary* activity. The black lines represent male time and the grey lines represent female time. In households with children, the number is set at the modal Australian category of two. The X represents the time of childless people, both men and women, in productive activity both as primary only, and as either primary or secondary activity.

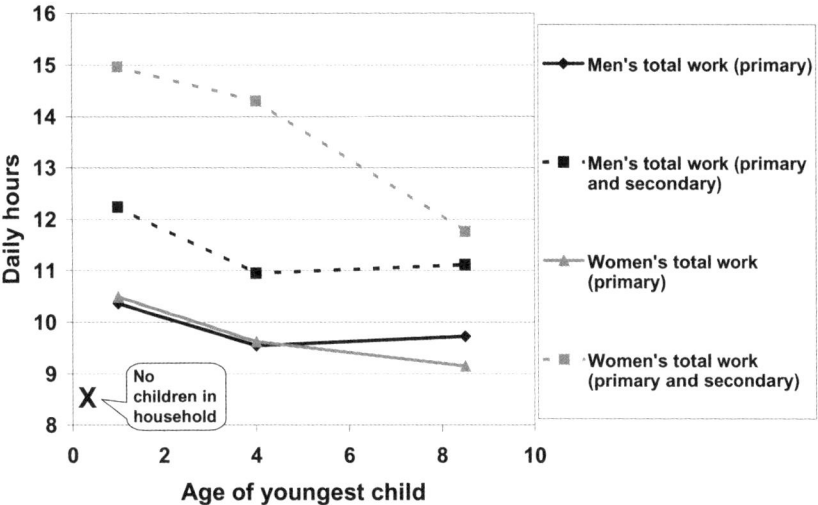

Figure 4.1 **Predicted hours a day spent by men and women in couple families in total productive activity (primary and secondary activity)**

Source: This graph shows fitted values from Table A6.

Figure 4.1 illustrates several findings. First, it shows that male and female total work times spent in *primary* activity are quite similar overall. Second, it shows the magnitude of the underestimation of parental time commitment that results from analysing primary activity only. Third, it shows that total workloads of childless couples are broadly equal, whether secondary activity is acknowledged (largely because secondary activity is negligible in these households) but that in households with children broad gender equity in total workload does not pertain if secondary activity is included in the calculation. Fourth, it shows that secondary activity is not absorbed into a mother's workload by compensating reduction in other activities, but simply adds to the workload.

Taking each of these points in turn, Figure 4.1 shows, firstly, that when primary activity only is considered (shown as the solid lines), men and women are doing about the same amount of work. Men and women who are childless spend almost the same amount of time in productive activity. Children bring greater time commitments to both sexes, but alter the comparative gender situation only slightly. However, looking at the dotted lines, which take account of secondary activity as well as primary, shows both the powerful effect of parenthood on time commitment and its differential effects by gender.

Childless men and women not only spend about the same time in unpaid work as a primary activity, but also are broadly equal when secondary activity is included in the count. Indeed, childless people of either sex only do a tiny amount of work as a secondary activity. Thus, the differences between parents and non-parents in time commitment are much greater than those indicated if analysing primary activity only. Fathers of two children, one of whom is an infant, have a primary activity workload about 20 per cent higher than that of a childless man. If secondary activity is also counted, the father's workload is 32 per cent higher than that of his childless counterpart. The discrepancy in workload is even more marked between mothers and childless women. Mothers have a much higher time commitment than women without children. When primary activity only is counted, the margin between childless women and mothers of two in total workload is 22 per cent. When secondary activity is included the margin in productive activity between childless women and mothers is over 100 per cent. On this measure, female total workload is more than doubled following parenthood.

Second, it is apparent that excluding secondary activity grossly underestimates the amount of time parents spend in productive activity. The mother of two children, the youngest of whom is aged 0–2 years, spends 10½ hours a day in paid and unpaid work as primary activity. When secondary activity is included, she averages just less than 15 hours a day in productive activity. When a mother of two has a youngest child aged 3–4, her workload as a primary activity averages nine hours and 40 minutes; her workload including secondary activity averages 14 hours and 20 minutes. So taking no account of secondary activity underestimates the time commitment of a mother with pre-school children by between 30 per cent and 40 per cent. When her youngest child goes to school, total work as a primary activity takes up nine hours and ten minutes a day of a mother's time. When secondary activity is included, she has a workload that absorbs 11¾ hours a day. This means that calculations that exclude secondary activity would underestimate her workload by 25 per cent. Fathers also do more work as a secondary activity than childless men, but the amount, both in total and as a proportion of their primary activity, is much less. Fathers of two spend about an hour and three-quarters a day more in work activities when secondary activity is included than if primary activity alone is counted. For men, there is little variation with the age of the youngest child; in all cases, excluding secondary activity undercounts male time in productive activity by 15 per cent.

Third, when secondary activity is included in the count of total productive activity, mothers' time commitment is much higher than fathers'. The total amount of work is not the same for each sex. Figure 4.1 shows clearly that the approximate equity found with primary activity does not hold when secondary activity is included in the analysis. On average, mothers are working 20 to 25 per cent longer than fathers when there is a pre-school child in the family, though this is reduced to nearer six per cent when the youngest child is at primary school.

Fourthly, the simultaneous unpaid work activity that a woman undertakes following the transition to motherhood is not absorbed into her total workload, but added to it. This is where the second shift can be found. Robinson and Godbey claimed that the second shift is a fallacy. Like other time-use researchers who find broad gender equity in total work time, they did not include secondary activity in their analysis (Robinson and Godbey, 1997; Gershuny, 2000; Beaujot, 2001; Bianchi et al., 2006). Measuring secondary activity as well as primary shows that this conclusion is only correct with regard to people who have no children. Broad gender equity in total workload pertains in childless couples, but it does not survive parenthood. Parenthood, for either sex, brings with it a requirement for secondary activity that is negligible for non-parents. This analysis shows that the second shift is not a fallacy, nor to be only metaphorically true in the sense that women are responsible for most unpaid labour. There is a more literal aspect too. The second shift does exist, but is restricted to parents, and mothers in particular. Further, it is contained in multitasking rather than in serial activity. The metaphor of a dual burden is quite apt. The second shift is being carried out not sequentially, but coterminously.

Gender Comparison of Relative Care Time

The gender discrepancy in absolute child care time is incontrovertible, but there is disagreement about whether this has implications for how mothers and fathers spend the time that they are with their children. Do men and women perform child care similarly in relative terms?

Child care task allocation by sex

There are also gender differences in how child care time is spent in relative terms. Table 4.1 shows the proportion of their own total child care time that mothers and fathers spend in each of the child care subcategories set out in Chapter 2. The child care activity subject to the most extreme difference in relative task allocation by sex is physical care. As a primary activity it accounts for over half a woman's child care time, but less than a third of a man's. Conversely, women spend 22 per cent and men spend 40 per cent of their time with children in interactive care. The parts of child care that are arguably most demanding (physical care) and the parts of child care that are arguably most pleasant (interactive care) are thus both unequally allocated by sex.

The child care tasks in which men mostly engage not only are arguably the most fun, but are also those that do not need to be done to a timetable. Men have more discretion over when they perform child care. Thus, women are not only more constrained by the amount of time they are undertaking care of children as a secondary activity, but also they are constrained by being the ones who perform the child care tasks that must take place at a certain time.

Table 4.1   Percentage of mean child care time spent by mothers and fathers in couple families in each child care task

	Primary Activity		*Primary and Secondary Activity*	
	Father	*Mother*	*Father*	*Mother*
Childcare Activity Category				
Interactive care	40	22	30	25
Physical care	31	51	13	21
Travel/communication	13	17	05	07
Passive care	16	10	52	47
Total	100	100	100	100

Source: Author's calculations of ABS TUS 1997.

Table 4.2   Percentage of mean child care time spent by mothers and fathers in couple families as primary activity

	Father	Mother
Childcare Activity Category		
Interactive care	49	34
Physical care	89	90
Travel/communication	96	97
Passive care	11	08

Source: Author's calculations of ABS TUS 1997.

Second, Table 4.1 shows the proportion of time allocated to each child care activity, whether it is done as a primary or a secondary activity. For both sexes, nearly half of their child care is supervision. However, for men, interactive care is the next largest category, as it was for primary activity only. The change is in the proportion of time women spend in physical versus interactive care. When secondary activity is included, women's time in talking and playing outweighs time in physical care. Table 4.2 shows how this is possible.

Simultaneous child care tasks by sex

Table 4.2 shows the proportion of time that each child care activity group is conducted as a primary activity. It shows a gender difference in the interactive care activities of talking to, playing with, reading to, teaching or reprimanding children. Half of male time in interactive care is done as a main activity. For mothers, only a third of the time spent in these activities is not done at the same time as other tasks. So when men play with or talk to their children it is often the only thing they are doing. Women more often do it at the same time as other activities. Through multitasking, they maintain time in these activities.

For both sexes, some child care activities are more likely than others to be done as a main activity (see Table 4.2). Physical care is much more likely to be recorded as a primary than as a secondary activity. Travel and communication is always coded as a primary activity. For other activities, secondary activity is more common. About 10 per cent of passive child care is recorded as a primary activity, with mothers spending slightly less of their child care time supervising children as a main activity than fathers. For both sexes, slightly more than a third of total child care is done as a primary activity. So while women spend more time in child care than men do in total, the proportion of that child care done as a primary activity does not differ much on the basis of sex.

Comparison of leisure quality by sex

The relaxing quality of leisure is diminished by the necessity to simultaneously supervise young children (Mattingly and Bianchi, 2003; Bittman and Wajcman, 2004). When children are under five, parents of both sexes spend a large part of their leisure with children present. However, there are gender differences. Not only do

Table 4.3 **Percentage of mean time spent by mothers and fathers in couple families by company present**

	Proportion of ...			
	...time with children that is childcare	*...childcare in sole charge*	*... all time with children in sole charge*	*... leisure not supervising children*
Father	29	13	08	48
Mother	50	33	29	40

Source: Author's calculations of TUS 1997.

fathers of young children average more total time in leisure away from children than mothers do (see Table A11) but also, even when children are with them, their leisure is less impinged upon than mothers in relative terms (see Table 4.3). During periods of recreation mothers are monitoring children for a substantially higher proportion of the time than are fathers. This implies fathers are not 'on duty' while with children as much as mothers are.

Responsibility for the job of child care by sex

The proportion of total time with children, and of performing child care tasks, respectively, that parents of each sex spend in sole charge of their children are shown in Table 4.3.

A smaller proportion of male time engaged in child care is spent in sole charge of their children than is female time engaged in child care. On average, slightly less than a third of male child care time is without their spouse present. In contrast, a woman with a youngest child under school age spends nearly half of the time she cares for children in sole charge. The sex discrepancy is more marked when not just child care time, but all time spent in the company of children is counted. Women spend nearly half of the time they are with their children in sole charge. Men are without their spouse present for about 15 per cent of the total time they spend with their children. This means that fathers are not substituting for their wives' time, and also that male time with children is overwhelmingly mediated by the presence of the mothers. Thus, men are not relieving women of responsibility for child care. Also, opportunities for men to develop independent relationships with their children are limited.

Sole Parents

As discussed in Chapter 3, sole parents must provide care to their children without the assistance of a resident partner. Does this mean that their children receive less care overall, or receive qualitatively different care from children in couple families? This section investigates how sole mothers compare to partnered parents, both mothers and fathers, in the amount and type of care they provide their children. Recall there are too few sole fathers to include in the analysis.

Table 4.4   Hours a day spent in child-related time by partnered parents and sole mothers

	Partnered Mother		*Lone Mother*		*Partnered Father*	
Childcare subcategories						
Physical care	1.59	***	1.64		0.20	
Interactive care	1.36	***	1.72	*	0.58	***
Travel for/with children	0.08		0.30	**	0.03	
Passive childcare	3.41	***	4.30	*	0.73	**
Company variables						
Hours a day in the company of children	12.32	***	13.07		6.83	***
Hours a day alone in the company of children	3.89	***	9.62	***	0.74	***
Child-free recreation	0.09		0.41	***	0.61	***

Source: Author's calculations of ABS TUS 1997. The figures in this table are drawn from Tables A11 (women) and A13 (men).

There is no significant difference in the amount of time mothers are engaged in primary activity care of their children according to whether or not they have a partner (see Table A10). Calculating fitted values for reference category respondents with two children suggests that both couple and sole mothers average just less than three hours a day in child care when it is counted as a main activity only. Equivalent men in partnered families average just over 50 minutes a day in primary child care. This may seem to indicate that children in couple families, having regard to fathers' input, receive more care in total. But it must be remembered that child care time is much more likely to be done as a secondary than a primary activity. When child care is calculated with both primary and non-overlapping secondary activity included there *is* a difference according to mothers' partnership status in the time spent caring for their children. When simultaneous activities are acknowledged, sole mothers of two average over eight hours a day in child care, compared to a partnered mother-of-two's average of six and three-quarter hours a day. Fathers in two-parent two child families average a total of an hour and 20 minutes in child care a day when both primary and secondary activity is counted. This is almost exactly equal to the difference between the time sole mothers and partnered mothers spend in child care. This suggests that sole mothers, through increased double activity, compensate in quantitative terms for the absence of a partner.

A finer comparison of lone mothers' and couple parents' child care time shows in which particular activities (physical care, interactive care (talking, reading, playing, reprimanding); minding children; and travel and communication with or for children) lone mothers make up for the absence of a resident father.

There is no statistical difference in the amount of time women spend physically caring for their children according to whether or not they have a partner. Both groups of women average just over an hour and a half a day. Partnered fathers average just less than eight minutes a day in physical child care. There is also no difference in the amount of time mothers with partners and mothers without partners spend in interactive care of their children as a primary activity. However, when secondary activity is acknowledged, sole mothers spend over 20 minutes more in interactive care of their children than women with partners do (see Table 4.4). This is about ten minutes less than the half hour that fathers in couples devote to this activity. Sole mothers, through double activity, manage to make up for more than half of the average interactive child care input of fathers.

Sole mothers allocate nearly three times as long, 14 minutes a day, to child-related travel than mothers with partners do. Australian sole mothers rely more heavily on care outside the household to substitute for their own care of their children than do women in couples. They are also more likely than couple mothers to use a mixture of care arrangements (Bittman et al., 2004). Therefore, this higher travel time may be a result of taking the children to and from these multiple arrangements. The largest time difference between how mothers with partners and mothers without partners spend their child care is in the sub-category of passive child care. Sole mothers average 53 minutes a day longer supervising their children than married mothers do, more than matching partnered fathers' time (48 minutes a day) in this activity.

So sole mothers spend more time in child-related travel, interactive care as a secondary activity and supervision of children than mothers with partners, and largely compensate for the absence of resident fathers. As shown above, fathers' child care activities are mostly conducted contemporaneously with mothers' care. So children in two- and one-parent families receive a similar daily duration of parental attention, but in couple families this includes time in which children enjoy the simultaneous attention of both parents. This comparison of time spent by lone mothers and the combined time spent by parents in couple families does not bear out concerns that children of Australian sole mothers are receiving much less time in each of the activities that comprise child care, or fewer total hours of parental care, than are children in two-parent families.

Time in the company of children

This is shown even more clearly when the variable considered is time spent alone in the company of children. Partnered and sole mothers spend the same amount of time in the company of their children (see Table 4.4). Both average about 12½ hours a day with their kids. Partnered fathers spend rather more than half that time with their children, about six and a half hours a day. However, there are big differences in the amount of that time during which they are the only adult present. Partnered fathers average about 45 minutes a day alone with their children. Partnered mothers average three and three-quarter hours a day alone with kids; sole mothers over nine hours. It is clear that a major consequence of not having a partner is that a sole mother is much more often alone with her children than a mother in a couple-headed family. Therefore, the time of a sole mother is more constrained than that of a partnered mother. In ensuring her children are supervised for the same time as those in couple families, the sole mother is afforded less opportunity than partnered mothers to undertake other activities such as paid work.

Child-free recreation

Perhaps mitigating the constraint identified above, sole mothers enjoy more recreation without their children present than do partnered mothers. Partnered mothers with a pre-school child average about five minutes a day child-free leisure. In contrast, sole mothers average nearly 20 minutes a day (see Table A11). A possible explanation is that their greater use of child care affords them the extra leisure. Also possible is that it reflects the input of non-resident fathers. The TUS time-diaries are recorded on only two days, and it is therefore a matter of chance whether those days will coincide with paternal access, but in some cases it will. When non-custodial fathers see their children, it is not usually in the mother's company. Thus, unlike the case for partnered mothers, in those families in which a non-custodial father maintains contact with his children, sole mothers have quite substantial periods of time in which they are completely relieved of direct child care responsibilities.

Discussion and Conclusion

In households with children there are profound gender differences both in total workload and in the way child care is conducted. However, these differences are hidden. Not only is the time and effort involved in unpaid labour invisible to usual economic measures (Waring, 1988; Folbre, 1991; Ironmonger, 1996; Gershuny and Sullivan, 1998) but mothers have a dual burden that is doubly invisible because it is also not apparent to usual time-use enquiry. This has allowed some time-use researchers to draw the conclusion that men and women, though continuing to specialise, have broadly equal overall time commitments to work in aggregate. Robinson and Godbey (1997) suggest that time-deepening behaviours were adopted in response to a perception of time stress rather than to the actuality of longer work hours, arguing that pace of life has quickened, but overall work time has not increased. However, this view is based on analysis of primary activity only. An analysis that also includes a measure of the time spent in unpaid work as a secondary activity gives a fuller picture.

The time-deepening behaviour of doing more than one thing at once is overwhelmingly the behaviour of parents. Parenthood, for either sex, brings with it a requirement for doing two things at once that is negligible for non-parents. For childless people, the difference between unpaid work done as a primary activity and when secondary activity is included in the count is minimal. Including secondary activity in the count of child care shows how hidden the time commitment associated with children is. It has been possible to 'prove' that the work times of each sex are equivalent, but only by overlooking an important part of the time cost of parenthood and the way it impacts differently upon men and women. Secondary activity does not take up more primary time, but requires more double tasks, and spills over into time that is ostensibly leisure. The analysis above shows that the contention that there is overall gender equivalence in work time is borne out only if solely primary activity is measured. A measurement that includes secondary activity shows that gender equity in overall work time does not hold in families with children.

My second focus of analysis in this chapter was whether, in addition to differences in total child care time, there are differences in the way time with children is spent and/or experienced by each sex. I looked at how androgynous male and female parenting practices are in relative terms. This snapshot of current practice shows there are substantial differences in the way men and women parent. Mothers do more interactive care than fathers, but it is a lower proportion of their total time in child care. Fathers, therefore, enjoy relatively more play and talking time with their children than mothers do. Mothers do more physical care than fathers in both absolute and relative terms. The child care tasks in which men mostly engage are arguably the more fun, which means that paternal time with children is less like work than is maternal time. This is not only important in terms of domestic equity, but arguably contributes to the perception of child care as more of a leisure activity than a work commitment, which, as argued in Chapter 1, is the neoclassical economic view that informs some social policy and work place attitudes to parents.

The tasks that men disproportionately perform are also those that do not need to be done to a timetable, therefore men have more discretion over when they perform child care. Women are not only more constrained by the amount of time they are doing child care; they are also constrained by being the ones who perform the child care tasks that must take place at a certain time. This is both cause and consequence of the situation in which men fit their family responsibilities around paid work, while women more commonly fit paid work around their family responsibilities (Blau et al., 1998).

Mothers are more likely than fathers to be doing two things at once while with children, and by doing more things at once they maintain time in certain activities, particularly interactive care. Mothers preserve time interacting with their children by multi-tasking, and thus working harder, than fathers. Mothers spend relatively more of their leisure time accompanied by children than fathers do. They are more likely to have their time with children interrupted to do other things. These findings mean that the job of child care is more intensive and fragmented for women, and that women have less recuperation time.

Married fathers are rarely alone with their children. This has two main consequences. First, it means that most male time with children is mediated by the mother's presence. When men perform child care, they are typically not substituting their time for their wives', but joining them as helpers in the task. If fathers are rarely alone with their children, they are not forging independent bonds with their children unmediated by the presence of the mother. Inter alia, this may have implications for the quality of father-child contact following divorce or separation (Furstenberg and Cherlin, 1991; Seltzer and Brandreth, 1994; Silverstein and Auerbach, 1999).

Second, if men are only helping women in the job of child care, it means that they are not equal participants. They fail to relieve women of responsibility for child care, including the aspect that many report to be the most onerous of all, that of managing the organisation and planning of care (Coltrane, 2000). Also, men are not participating in child care in a way that fully substitutes for female time, even proportionally. This dashes expectations that male involvement in child care would relieve women of their domestic responsibilities and free them to pursue other activities such as paid work. Men are not undertaking child care in a way that relieves women of the responsibility for care and substitutes for female time, even proportionally.

Because fathers' contribution so often overlaps that of their wives possibly explains why sole mothers manage almost to match couple families' levels of time with children. Sole mothers do not spend a different amount of daily time in the company of their children than couple mothers. What does differ is the amount of time that sole mothers are *alone* with their children. Lone mothers are not working longer or harder than mothers in couple families, but they are more constrained and burdened by sole responsibility associated with being unable to leave their children. It seems likely this is a major reason that sole mothers in Australia spend even less time in the paid work force than mothers with partners. However, on the upside, children of sole mothers do not miss out on much parental care and company, and sole mothers enjoy more leisure without their children present than do couple mothers.

In this chapter, I investigated the accuracy of two views of the gender distribution of workload in households with children. One approach sees decreasing specialisation in types of productive activity on the basis of sex. Proponents also contend that currently there is broad equality between men and women in the time spent in total productive activity and leisure and that there is no second shift (Robinson and Godbey, 1997; Gershuny, 2000; Beaujot, 2001). The other view is that sex continues to be the major factor in domestic task distribution, in that women are adding employment to their productive activities but that men are not adding housework and child care to theirs. Wives are therefore working much more than their husbands, in order to meet the demands of this dual burden (Hochschild and Machung, 1989; Schor, 1991).

I find that the latter view is ultimately the more accurate. Time-use analysis offers the opportunity to make visible the huge amount of time and effort involved in unpaid labour. The conclusions of the first approach result from using time-use analysis in a narrow way, which obscures recognition of important aspects of 'life as it is lived' (Gershuny and Sullivan, 1998). Including secondary activity in the analysis shows that in households with children, mothers have a much more substantial time commitment to work activities than fathers do. That this time is doubly invisible, both to standard economic enquiry and to shallow time-use calculations, is a major reason why the constricting impact of becoming a mother is not widely recognised. It also explains why each side of the argument thought that the other misperceived the problem, the first holding that the disparity between male and female workloads is exaggerated, and the second that it is underestimated and minimised. This analysis shows the limitations of using time use data to measure what is easiest – the overall time spent in each task as a main or 'primary' activity. While such an approach allows the counting of important aspects of life that have been absent from economic measures, it is still a sex-biased measure (Sullivan, 1997; Craig, 2006b; Bryson, forthcoming; Craig, 2007). Simply adding up time spent in each activity, though useful, is not adequate to a full recognition of the impacts of children.

Second, even in relative terms there are significant differences in the way fathers and mothers parent. The experience of caring for or being with children is not the same for each gender. This has implications for quality of life for the parent, quality of care for the children, for constraints upon sole mothers and for women's participation in paid work, but the difference is subtle enough to be difficult to identify and make obvious. Both the amount of work involved and the differences in the way each sex conducts child care are obscure, possibly even to mothers and fathers themselves.

In the next chapter I investigate the effects on parents' time when they use substitute, non-parental care for their children.

Chapter 5

Protecting Parental Time with Kids?[1]

Mothers spend much more time caring for children than fathers do, and also do more of the regular and demanding aspects of care, as previous chapters have shown. This chapter looks at what happens to parents' time when they use substitute, non-parental care for their children. In some quarters there is concern that as women increasingly participate in paid work, children will receive insufficient care. This fear is based on the assumption that to spend time in market work parents must, unavoidably, reduce their hours of contact with their own children and diminish the quality of the parenting they provide. However, emerging evidence raises questions about whether there is much difference in maternal time spent with children by working mothers than by non-working mothers, particularly in active, involved care. This chapter presents the findings of studies that show there is not a one-to-one trade-off between parental and non-parental care, and goes on to investigate how this is possible. How do mothers who commit time to the paid work force, or use non-parental care to substitute for their own care, maintain a similar amount of time with children as non-working mothers?

Background

Mothers are entering the workforce in increasing numbers across the western world. Participation rates have historically been high in Scandinavian countries, and over the last 50 years have steadily grown elsewhere. Half of Australian mothers in two-parent families are working by the time their youngest child is 1 or 2. The 'traditional' father-breadwinner, mother-homemaker family represents just 27.5 per cent of families with children under 5 years old. Only 18.1 per cent of families with children under 8 years conform to the stereotype (ABS Census, 2000). In the US, maternal employment has tripled over the past 30 years (Spain and Bianchi, 1996). In 1997, 63.9 per cent of women with children under 6 and 78.3 per cent of women with children aged 6–17 were employed (Perry-Jenkins et al., 2000). The trend includes women with very young children. In 1994, almost 60 per cent of US mothers with children under 3 were employed, compared with 21.2 per cent in 1966 (Blau et al., 1998). It is projected that by 2010, female workers will account for 47.9 per cent of the working population in the US (NIOSH, 2004).

---

1   Part of the argument in this chapter, and Figures 5.1 and 5.2, appear in Craig, Lyn (2007) 'How employed mothers in Australia find time for both market work and childcare' *Journal of Family and Economic Issues*, 28 (1): 69–85. The original publication is available at www.springerlink.com.

Maternal workforce participation is facilitated by non-parental child care. The provision of good quality institutional child care was seen by feminist reformers as an essential prerequisite to women's freedom to earn a living (Bergmann, 1986; Brennan, 1998). However, institutional care does not have universal approval as a method of balancing work and family, because many assume that working mothers must inevitably deprive their children of substantial maternal care. Outsourcing care is presumed to facilitate women's adoption of male work-care patterns, which would logically mean that children would receive less parental care. Those who value the idea of parents personally caring for their own children find female withdrawal from child care problematic (Pfau-Effinger, 2000; Gornick and Meyers, 2004). As the trend to maternal workforce participation grew throughout the western world, so did the concern that as a result children would not be adequately cared for (Hochschild, 1997; Hewlett and West, 1998). Despite working motherhood now being the statistical norm, there is continued unease over the consequences for children (Presser, 1995; Arundell, 2000; Gornick and Meyers, 2003; Leach et al., 2005). This disquiet is underpinned by theories of child development and psychology, developed over the last century, which suggest that maternal bonding, attentive parenting and high time inputs are necessary for optimal educational and social outcomes for children (Bowlby, 1972; Bowlby, 1973; Belsky, 2001). If these precepts are accepted, mothers are faced with a choice between economic independence and providing optimum care for their children. The wish or need of women to work and the belief that children require the full-time presence of a mother, are incompatible.

The issue resonates at a social level. Attitudes to out-of-home child care are determined in part by what people think about women's right to autonomy and in part by what they think is good for children (Bergmann, 1986; Brennan, 1998). Social ambivalence about maternal employment can be reflected in a lack of policy support for the use of non-parental day care. The cost, quality and supply of non-parental care are influenced, both directly and indirectly, by public policies. The relatively high female workforce participation in Scandinavia is widely attributed to the public provision of accessible and affordable child care, and the well-established social and political acceptance of care delegation (O'Connor et al., 1999). A political view that children are best served by full-time mother-care can find expression in policies that lead to inadequate or expensive day care services (Pocock, 2003a).

In Australia in 2002, about half of all children under 12 years old used some form of child care, either formal or informal.[2] Formal care was used by about a

---

2   In Australia, 'formal child care' refers to regulated care away from the child's home. It includes before and after school care centres, long day care centres, family day care (in which registered providers care for up to five pre-school children in their own homes), nursery school and kindergarten centres, and occasional care centres. "Informal child care" refers to non-regulated care in either the child's home or elsewhere. Informal care includes care provided by the child's siblings, grandparents, another relative of the child, or any other person (ABS, 1998, *Time Use Survey, Australia. Users Guide 1997 Cat No. 4150*. Canberra: Australian Bureau of Statistics). It may be paid or unpaid. Australian regulations require monitoring of the quality of formal childcare and the enforcement of standards. The Quality Improvement and Accreditation Scheme licenses long day care centres for one-, two-, or three-year periods according to their compliance with 52 criteria. The Federal Government makes Child

quarter of Australian children under 12. Around a third of children used informal care, with a large majority of these (19 per cent of all children) being cared for by their grandparents. About a third of Australian children were in a mixture of formal and informal arrangements (ABS, 2002). The Australian formal child care regulations are relatively demanding by world standards and therefore formal child care is generally of high quality (Brennan, 1998), but can be both hard to access and very expensive (Pocock, 2003a). There are long waiting lists, particularly for the very limited places available for under-3-year-olds (Castles, 2004). Recent figures from the ABS suggest that lack of affordable and accessible child care is a significant barrier to female workforce participation, which in 2004–2005 prevented over 250,000 women who wanted a job, or who wanted to work more hours, from doing so (ABS, 2006).

Social policy in Australia tends to reinforce traditional gender roles (Forssen and Hakovirta, 2000; Charlesworth et al., 2002; McDonald, 2004b). Many Australians do subscribe to the view that young children require full-time maternal care (Evans and Kelley, 2002). In such a milieu, concern that employed mothers are depriving their children of vital maternal care may be shared by working mothers themselves, who may feel ambivalence and guilt at leaving their children in the care of others (Arundell, 2000).

In response to the concern over child welfare, there has been a great deal of research into the effect of non-parental care on child development and outcomes. The results are mixed, but do not indicate that non-parental care is necessarily harmful to children (Presser, 1995; Bianchi and Robinson, 1997; Han et al., 2001; Zick et al., 2001; Leach et al., 2005). The effects of non-parental care vary with the age of the child. Some negative effects on behavioural and cognitive outcomes have been found if children attend long day care when under 1 year old, whereas older children have been found to benefit (Hoffman and Youngblad, 1999; Belsky, 2001; Han et al., 2001; Brooks-Gunn et al., 2002). A recent longitudinal study of British children found that for babies and toddlers up to 18 months, group day care is not as good as one-on-one care. They found that children looked after at home by their mothers showed better social and emotional development than children who had been in day care. They found that other home-based care, including grandparents, childminders and nannies were (in ascending order) better than day care (Leach et al., 2005).

However, negative effects are mediated by other factors including the characteristics of the child (for example, temperament), the characteristics of the family (for example, income and parental education), and the quality of the day care institution itself (for example, having well-trained staff and high carer–child ratios) (NICHD, 1997; Blau, 2000; Shonkoff and Phillips, 2000; Belsky, 2001;

Care Assistance (CA) payments to parents who use accredited centres. State governments monitor preschools, kindergartens and occasional care centres. Family day care is also a state responsibility. Local groups apply for a state government license to administer the service. They must demonstrate how they would address 156 items on a risk assessment list. Once licensed, the organisation assumes responsibility for recruiting, assessing and monitoring caregivers. The quality of informal childcare, which is provided privately by relatives, friends and neighbours, is unknown and there is no enforcement of minimum standards.

Han et al., 2001; Leach et al., 2005). The most important single mediator is the family environment (Shonkoff and Phillips, 2000). The US National Institute of Child Health and Human Development (NICHD) found that non-parental care does not have a detrimental effect on child outcomes unless poor quality care is combined with poor parenting (NICHD, 1997).

This suggests that parenting quality is not necessarily diminished concomitantly with the use of non-parental care. Also, it may be that the debate rests on inaccurate assumptions. Perhaps the picture on child outcomes is inconclusive partly because maternal employment hours and time in non-parental care are only very approximate indicators of parental time with children. It was assumed that paid work and time with children would be traded off against each other – women who worked or used non-parental care would necessarily lower the time they spent caring for their children. But emerging evidence suggests that the assumption that non-parental child care and maternal employment actually equate with a substantial loss of parental attention to children is misplaced.

Maintaining Time with Children

Apparently paradoxically, mothers' workforce participation and parents' rising use of child care centres has been accompanied by increases in the time both mothers and fathers spend in face-to-face activities with their children. A growing body of time use evidence now shows that mothers do not reduce the amount of time they spend with children by the same amount of time as they spend in paid work. Maternal child care is reduced by far less than an hour for every hour the mother works (Nock and Kingston, 1988; Bryant and Zick, 1996b; Bianchi, 2000; Hofferth, 2001; Sandberg and Hofferth, 2001; Booth et al., 2002; Bittman et al., 2004). Also, historically, parental time with children has not declined (Nock and Kingston, 1988; Hofferth, 2001; Sayer et al., 2004; Bianchi et al., 2006). Overall, time with children has not decreased alongside the increase in female employment. While the time children spend at home has decreased, the time that parents spend in activities with children has not (Bryant and Zick, 1996a; Bianchi, 2000; Bianchi et al., 2006). It appears that the impact of structural change in female employment practices upon time with children has been outweighed by behavioural change in time mothers spend with children (Sandberg and Hofferth, 2001; Sayer et al., 2004). Employed mothers do spend less time in activities with children than non-employed mothers in the cross-section (Nock and Kingston, 1988; Bryant and Zick, 1996a; Bryant and Zick, 1996b; Robinson and Godbey, 1997; Pocock, 2003b), but when mothers increase their hours of paid employment, they do not reduce time spent in activities with children by an equal amount (Bianchi, 2000). Studies have found that over a lifetime (depending on when they return to work), employed mothers spend between 80 and 90 per cent of the time in child care activities that stay-at-home mothers do (Bryant and Zick, 1996a; Bryant and Zick, 1996b; Sandberg and Hofferth, 2001). While these results should be interpreted with caution (Budig and Folbre, 2004), they clearly suggest that maternal employment has smaller effects than is widely assumed.

Further, the evidence suggests that employed mothers use the time that they are with their children to do almost as much of certain types of child care as non-working mothers do. They prioritise active, engaged child care time over non-engaged, supervisory time (Bittman et al., 2004). Nock and Kingston (1988) found that non-employed mothers spend more time with their children than employed mothers do, but the bulk of this additional time is not spent in direct interaction with the children. While employed mothers spent four hours a day in the company of preschoolers and non-employed mothers spent nine hours a day, the difference in time spent in direct interaction with preschoolers by employed and by non-employed mothers was less than an hour (Nock and Kingston, 1988). Similarly, using data from the Child Time Use Supplement of the US Panel Study of Income Dynamics (PSID) for 1997, Hofferth found that maternal employment had a significant effect on time that mothers were present, but not interacting with their children, but not on active participation in activities with children (Hofferth, 2001). These findings imply that parents in general, and mothers in particular, may target a certain minimum level of time with their children, and make other adjustments in order to meet that target. Employed mothers seem to prioritise high quality time with their children.

Most discussions of child outcomes centre on the effects of non-parental child care. However, the time-use studies above have investigated the effects of employment, rather than child care itself. There is a widespread assumption that the two are interchangeable measures, in that the residual of either will be time available to spend with children. With few exceptions (Booth et al., 2002; Bittman et al., 2004; Craig et al., 2006) studies have not explored how non-parental care usage specifically, affects quantity and quality of parental time with children.

Booth et al. (2002) conducted a small-scale qualitative study which compared the time-use of 143 mothers of 0–6-month-old babies in families in which the infants spent more than 30 hours a week in day care, with the time-use of 183 mothers whose infants spent no time in day care. They categorised types of care into instrumental care and social interaction. Consistent with the studies above they found that the difference in interaction time was less than expected. 'In care' mothers spent about 12 fewer hours than 'no-care' mothers interacting with their babies, which was less than half the 30-plus hours the children spent in care. In addition, they found that the quality of mother-child interaction did not seem affected by the use of day care, and that fathers in 'in-care' families were more involved in care-giving (Booth et al., 2002).

In a quantitative analysis using TUS data as described in this book Bittman, Craig and Folbre (2004) investigated how the use of non-parental care affects the quantity and quality of parental time in activities with children. They examined the relationship between the use of non-parental care and parental time in activities with children, both in overall amount and in time allocated to the four-way typology of parental care activities (interactive, physical, travel and communication associated with children, and passive care) described in Chapter 2. They found no type of parental child care to be reduced on an hour-for-hour basis by the use of non-parental child care. Some were not reduced at all.

In particular, parents defended the time they spend in interactive (talk-based) activity with children. Such activities as talking, playing and reading with children, thought to be the most essential in enhancing children's cognitive and social development (Brooks-Gunn et al., 2002), did not significantly vary with either the type nor the quantity of non-parental care utilised. Bittman et al. found that children enjoy the same amount of time in interactive activities with their parents, regardless of the weekly hours children spend in care outside the home, or the type of care used. This type of parent-child interaction was neither increased nor reduced in association with institutional care use. They did find, however, that non-parental care replaced some of mothers' time in physical care of their children. Mothers who used care spent less time in physical activities (such as feeding and changing diapers) than did mothers who used no non-parental care. This may be because physical care is less amenable to rescheduling than the talking and playing activities that comprise interactive care. Children have to be fed when they are hungry, cuddled when they are crying. However, other physical care such as bathing a child, or putting a child to bed, are likely to be performed outside the hours of a traditional paid workday, and are therefore still likely to be done by parents. So, perhaps for this reason, non-parental care does not substitute for some physical maternal care on an hour-for-hour basis. Mothers who use non-parental care delegate between a third (informal care users) and a fifth (formal care users) of it, meaning that even mothers who find substitutes for time with their children retain the majority of hands-on, physical care.

Perhaps because fathers do much less of the physical care of children (the average was only 12 minutes per day) than mothers, Bittman et al. found their time in physical care activities with children to be completely unaffected by use of non-parental care, and by their paid work hours. Nor was fathers' allocation of time to child-related travel and communication affected by the use of extra-household child care. In contrast, mothers' child-related travel and communication went up (by 12 minutes a day), but only if a mixture of formal care and informal care was used. There was no effect if only one type of care was used, and the result is probably because a multiplicity of care arrangements requires more travel between venues, and more communication with others.

Consistent with the research discussed above, Bittman et al. found that the largest impact of non-parental care and time spent in paid work on any aspect of a parent's time with children was on mothers' passive, supervisory care. The longer children spent in care, the less passive care the mother did. This suggests that women who allocate time away from their children alter the composition of their own care, which implies that mothers protect time in the most actively involved forms of child care. This casts doubt on the supposition that placing the care of children in the hands of others denies children the most important forms of contact with their parents. Mothers who use non-parental child care spend proportionally more of their own child care time physically caring for their children, or in verbally interacting with them, than do mothers who do not use substitute care. They preserve 'quality time' with their children, that is, time in the interactively most significant types of parent contact. Non-parental care substitutes for low intensity activities, and a higher proportion of time that mothers do spend with children is spent in interactive and physical activities.

The gender division of labour in activities with children becomes somewhat more equal in households that use non-parental care. Consistent with the results in Chapter 3 of this book, Bittman et al. found that the expenditure of time by mothers in nearly all activities with children is three to four times greater than that of fathers. Interactive activities are an exception to this pattern, since mothers and fathers devote nearly the same amount of time to these (as a result, it comprises a much greater proportion of fathers' than mothers' total time in activities with children). Fathers spend less time than mothers in activities with children, and their time inputs vary considerably less with family circumstance. Therefore, the use of non-parental child care reduced mothers' time with children more than fathers'. So as hours of non-parental care increase, the gender division of labour in activities with children becomes more equal. Also, some maternal physical care is transferred to non-parental caregivers, thereby increasing the proportion, though not the quantity, of the total household physical care that is undertaken by fathers. The way mothers who use non-parental care allocate time to each type of parental care is somewhat closer to the way in which fathers allocate their care time. Non-parental care substitutes for low intensity activities, and a higher proportion of time that mothers do spend with children is spent in interactive care activities. Fathers' time in both interactive and physical child care activities does not much change with extra-household child care use. In contrast, the use of formal child care is associated with significant increases in paternal passive care time. So hours of care outside the home reduce mothers' time spent in physical child care activities, but increase fathers' involvement in passive care activities. Thus, the distribution of child care time between mothers and fathers becomes more equal as hours of non-parental care increase. This does not mean, however, that men and women who use non-parental care are doing the same type of child care as each other. While the provision of non-parental care allows mothers to delegate some of their routine responsibilities to paid care providers, and therefore interactive activities come to represent a larger proportion of the time they spend with children, the gendered child care task allocation identified in Chapter 4 of this book appears to largely persist. Men in households that use non-parental care are spending longer in children's company and fulfilling a supervisory role, but they are not doing more physical care.

In summary, parents use non-parental child care to substitute for themselves, but do not do so on an hour-for-hour basis. It seems that parents try to make up for the time the children are in care. These findings testify to the resilience and flexibility of parental commitments to childrearing. They suggest that mothers, in particular, prioritise certain types of activities with their children and rearrange their schedules to accommodate them. They further suggest that some activities cannot be delegated. As a result, increased hours of maternal employment and increased utilisation of non-parental care lead to only small reductions of parental time in activities with children. These results challenge the assumption that employment and non-parental care deny children vital parental care. Despite the huge concern, discussed above, that children might be missing out on essential care, there is actually less reduction in parental time than was widely anticipated. This suggests that the findings of psychological research that shows non-parental care is not inimical to good child outcomes (NICHD, 1997; Blau, 2000; Shonkoff and Phillips, 2000; Belsky, 2001;

Han et al., 2001), is arguably not only because substitute care may not in itself be harmful to kids, but also because they do not actually get much less parental interaction time.

The expectation that women must make a choice between work force participation and providing a high standard of care to their children has not been borne out. Women appear to resist such a trade-off. Full delegation is not apparently desired, and in the case of physical care particularly is probably not possible. The research implies that women, even those who allocate substantial time to market work, may target a certain minimum amount of interaction time with their children, and then seek ways of meeting that target (Bittman et al., 2004). Employed mothers make compositional changes in their time with children (Sandberg and Hofferth, 2001; Bittman et al., 2004), and preserve their time with children over the longer term (Cohen and Bianchi, 1999; Sayer et al., 2004). This seems to support the idea that social norms of involved motherhood have not been reconciled with the trend towards increased female work force participation. For women who wish to earn a living through market work and also feel a strong imperative to care intensively for their own children, a difficult friction point results. If women value both paid work and attentive parenting, they will be reluctant to trade off child care time for time in market work, and will instead try to retain both. This is perhaps reassuring from the perspective of child welfare. However it does raise an obvious question: how do they do it?

How do Working Mothers Maintain Time with Children?

How do mothers who undertake paid labour or place their children in non-parental care manage to spend substantially similar amounts of time in child care activities as non-employed mothers? If market work and parental child care are both prioritised, the logical corollary is that other forms of time use, that is, non-employment and non-child care activities, must be adjusted. Apart from doing more at once, time for children can be found by reducing time in other activities and directing it to child care time, or by rescheduling time with children around other activities.

The analysis in Chapter 3 showed that parents, particularly mothers, in households with children spend less time in sleep, personal care, recreation and leisure than adults in childless households. If employed mothers do not completely trade off market work and child care, the implication is that they need to reduce such activities even further than do mothers who are not employed. Employed mothers presumably maintain their time commitments to both paid work and child care by rescheduling (shifting) their child interaction time and their other time commitments around their market work. Since non-parental care does not simply relieve mothers of care on an hour for hour basis, it may be that non-parental child care is used to facilitate the shifting and rescheduling of parental child care time.

An assumption in much of the literature is that non-parental child care and maternal employment are interchangeable measures in that the residual of either will be time available to care for children. However, because non-parental child care is used for both work and non-work purposes, this is misleading (Bittman et al., 2004). Many mothers use non-parental care to do things other than paid work, and some work is undertaken with children present. Non-parental care is used not only to replace time that mothers are employed, but also time that mothers spend in other activities. Therefore, to assume that they are commensurate, or to rely on either as a proxy for time with children will yield noisy results. There is a possibility that widespread child care usage for non-work purposes and the practice of using no child care while employed may have obscured the possibility that in addition to replacing some parental care time, non-parental child care is used to facilitate the shifting and rescheduling of parental child care time.

I now explore the question of how mothers ensure that non-parental care does not diminish their care time on an hour for hour basis. I speculate that mothers use child care not to replace their own care, but to shift the times when they are together with their children. Non-parental child care is used for both work and non-work purposes. Non-working mothers have more flexibility than working mothers to reschedule their parental child care around substitute care. Therefore, working mothers preserve their time with children by reducing, in comparison with non-working women, the time they spend in non-work and non-child care activities, and rescheduling child care activities to later or earlier in the day.

Research Focus and Method

The analysis for this chapter is conducted in three stages. First, I conduct OLS regression analysis, using a sub-sample of the TUS (parents with at least one child under 5 years old) and model as described in Chapter 2, with the dependent variable the weekly hours of non-parental care used. The analysis is run separately for men and women. The intention is to see which demographic variables predict the use of non-parental care. I also give a brief descriptive overview of the extent to which, in this sample, maternal workforce participation and child care go hand in hand.

Second, to avoid the effects of child care usage for non-work purposes confounding the analysis, I split the women in my sample by workforce status and run the model separately for fathers, working mothers and non-working mothers. The model is the same as for the above analysis except that it includes dummy variables for the type of child care used, and because the women were separated by workforce status, the variable 'hours of market work' is excluded from the female regressions. The intention is to see whether the type or duration of non-parental care has different implications for each group. The dependent variables are four separate types of non-work and non-child care activity that may be sacrificed to child care activities: housework, sleep, personal care and child-free recreation time (as described in Chapter 2). The constant terms represent the amount of time spent in the dependent variables on a weekday by parents of one child under 2 who have no post-school education, use no non-parental child care, are aged 35–44 and have no disabled household member.

For the final part of the analysis, I calculate whether respondents are participating in active child care[3] in each 5-minute block of time during the 24-hour day. I then compare the average participation in active child care at each end of the day in households with mothers working full-time (35 hours a week or more) and households with mothers who do no paid work. This is intended to investigate whether in working mother households child care activities are rescheduled to earlier or later in the day than in other households.

Results

Who uses non-parental child care?

Mothers   Non-parental child care usage is predicted by mothers' employment (see Table 5.1). However, time in employment and time in child care do not equate. Non-parental care is predicted to go up by half an hour a week for every hour a week a mother works. For women, spouse's employment does not predict an increase in the use of non-parental child care. Being a single parent predicts a large increase in non-parental care of nearly 12 hours a week, and two children in the family predicts that non-parental child care will average 2.17 hours longer than if there is one child, or if there are three or more children in the family. The use of child care is also predicted by maternal age. Mothers aged 25–34 average 2.4 hours a week more non-parental child care usage than mothers who are older. There is a small but highly significant association with household income. Non-parental care increases by 0.005 hours (0.3 of a minute) a week for every extra dollar of household income.

Fathers   The predictors of non-parental child care for men are complementary to those for women. With regard to female employment, the findings for men mirror those for women. Whereas male time in paid employment does not predict greater non-parental child care use, men's spouses' hours do predict it. Non-parental child care goes up by nearly half an hour a week for every hour a man's spouse works. These results confirm that non-parental child care is used to replace mothers' time, not fathers'.

In findings that echo those for women, child care usage goes up by a tiny but significant amount with household income (+ 0.005 hours a week − 0.3 of a minute) for every weekly dollar earned. Non-parental child care is predicted to rise by 2.2 h a week if there are two children in the family.

Although female time in paid work strongly predicts the use of non-parental care, it does not perfectly equate with the amount of time children are actually in that care. Part of the mystery of why the use of non-parental child care does not markedly reduce mothers time with children is resolved by investigating the purposes for which child care is actually used.

---

3   This variable was created for this analysis from ABS activity codes 500–530 and 550–599. It includes all types of child care that are active (physical care, interactive care, child-related travel and communication), rather than supervisory performed as either a primary or secondary activity.

Table 5.1   Predictors of non-parental child care use

	Father		Mother	
Variable				
Sole parent	N/A		11.71	***
			(1.82)	
Child >2 years	0.71		1.45	
	(0.97)		(0.91)	
Number of children				
Two	2.20	**	2.17	**
	(1.07)		(1.00)	
Three or more	0.35		1.00	
	(1.18)		(1.10)	
Market work (hours a week)	-0.10		0.50	***
	(0.03)		(0.03)	
Spouse's market work (hours a week)	0.46		-0.01	
	(0.03)		(0.03)	
Household income	0.01	***	0.01	***
	(0.00)		(0.00)	
Disabled person in household	0.93		-0.41	
	(1.01)		(0.97)	
Aged				
25-34	0.21		2.50	**
	(0.93)		(0.93)	
45-54	-3.57		-9.21	
	(2.17)		(4.16)	
Qualifications				
University	-1.69		-0.27	
	(1.12)		(1.07)	
Vocational	-0.88		0.89	
	(1.05)		(1.04)	
Day of the week				
Saturday	1.06		0.01	
	(1.28)		(1.21)	
Sunday	1.67		1.45	
	(1.67)		(1.18)	
R Square	0.36		0.40	

* P-value<0.05 ** P-value<0.01 *** P-value<0.001 N=1690
Source: Author's calculations of ABS TUS 1997.

As discussed above, a widespread assumption is that non-parental care is used to allow a mother to spend time in the paid work force (Brennan, 1998). This is indeed the major use of extra-household day care, but it is by no means the only use for it. Non-parental care is used not only to replace time that mothers are working, but also time that mothers are spending in other activities. In my sample, paid work and the use of non-parental child care have a correlation of 0.47 for married mothers and 0.31

for sole mothers. Not only is non-parental child care being used for other purposes than employment but also, women are working but not using non-parental child care when they do so. A close examination of individual case records reveals that the discrepancy between women's hours of work and use of extra-household child care can be explained. There were 111 cases of women who were in the paid work force, but used no regular weekly non-parental care. Of these, there were 50 cases of women who had actually done paid work on the diary days, but used no non-parental care. Some were doing shift work, but the most usual situation was that the work was carried out at home with the children present. Most of the women who employed this strategy were working part time, doing clerical work for the private sector, although some worked in agriculture. An examination of their husbands' records suggested that some of the women were farmers' wives who participated in farm work while supervising children, and others were doing the clerical work for their husbands' business while supervising children. In my sample, 468 women (52% of the total number of women) did paid work. Of these, 166 (35%) used formal care only, 90 (19%) used informal care only, and 101 (22%) used mixed care. One hundred and ten (24%) did not use non-parental child care. Four hundred and twenty-one women (47% of the total number of women) did no paid work. Of these, 125 (30%) used formal care only, 21 (5%) used informal care only, and 25 (6%) used mixed care. Two hundred and fifty (59%) did not use non-parental child care.

Paid work and non-parental child care do not go hand in hand. Many women who work do not use non-parental child care and many women who use non-parental child care do not use it for work purposes. So part of the answer to the question of how women preserve time with their children despite non-parental child care use, is that both working and non-working women use non-parental child care, and it has different implications for each group. Women who use non-parental child care for non-work purposes can fit their parental child care around their non-parental care arrangements. Similarly, working women who do not use non-parental care can fit their work around their care responsibilities, or do both simultaneously.

However, this leaves the greater challenge, which is to explain how mothers who use non-parental child care in order to work maintain their parental child care time.

Table 5.2　　Proportion of households using non-parental care by mother's work-force status

	Women in Paid Work (N=468)	Women Not in Paid Work (N=421)
Type of non-parental care	%	%
None	24	59
Formal only	35	30
Informal only	19	5
Both formal and informal	22	6
Total	100	100

Source: Author's calculations of ABS TUS 1997.

I now turn to the question of where these women find the time to maintain their time inputs to children. The model is the same as for the analysis above, except that types of child care (formal, informal and mixed) have also been entered (see model specifications, Table A1). Demographic variables impact differently upon the time use of men and women. Similarly, working women and non-working women may be affected differently by different variables. Therefore, the groups are separated for this analysis.

Domestic Labour

One source of the time devoted by working mothers to care of their own children is time that non-working women devote to domestic labour, that is, housework and shopping (see Table 5.3). Note the constant terms – referent group working women spend 3 hours and 12 minutes a day doing domestic work and shopping, compared with nearly 5 hours spent by non-working mothers. Fathers with a similar demographic profile average much less time than either group of women (1.9 hours a day) in housework and shopping.

The model predicts that the use of non-parental child care will further reduce working mothers' time in unpaid domestic work such as housework and shopping. Working mothers' time in domestic labour reduces by 1.2 minutes a day in association with every weekly hour of non-parental care (amounting to over half an hour a day for 30 hours of care a week). It is unaffected by the type of care used. Mothers who do no paid work but use mixed care are predicted to also spend less time in domestic labour than mothers in the reference category, by 0.4 hours a day.

Working mothers' time in domestic labour is predicted to go up with each extra hour of paid work done by their spouse. This amounts to nearly an hour a day if he works a standard 35-hour week. Both fathers and working mothers are predicted to catch up on domestic duties on the weekends. Men spend about an hour and a half longer in domestic chores on weekends than on weekdays. The model predicts that working mothers will do nearly an hour more housework on a Sunday than on a weekday. Non-working mothers do not appear to reschedule like this, and average no more housework at the weekends than on weekdays.

Sleep

Parents get less sleep than non-parents, and mothers lose more sleep than fathers (see Chapter 3). The average sleep time of mothers who work and mothers who do not is fairly similar (see Table 5.4). Fathers in the base category of the regression model average about 25 minutes more sleep a night than either working or non-working mothers. Using non-parental care seems to gain working mothers a little extra sleep. Duration of non-parental child care is associated with a small but significant increase in sleep time for working mothers. The predicted increase would amount to about a day if the child were in day care for 20 hours a week.

On average, all parents get some extra sleep on a Sunday. Fathers average an hour and six minutes more, working mothers 42 minutes more, and non-working

Table 5.3   Coefficients of hours a day spent in domestic labour

	Domestic labour		
	Father	Mother	
		Employed	Not Employed
Variable			
Constant	1.99 ***	3.24 ***	4.98 ***
Type of non-parental care			
Mixed	-0.28	-0.66	-0.46 *
Formal only	-0.21	-0.33	0.18
Informal only	-0.03	-0.68	-0.31
Duration non-parental care (hours a week)	-0.00	-0.02 *	-0.00
Market work (hours a week)	-0.10 **	N/A	N/A
Spouse's market work (hours a week)	0.00	0.00 ***	0.00
Household income	-0.00	-0.00 *	-0.00
Single parent	N/A	1.05	0.47
Child >2 years	-0.07	0.17	0.53
Number of children			
Two	0.12	0.37	0.08
Three or more	0.08	0.53	0.69 *
Disabled person in household	0.47 **	0.16	-0.01
Age			
25-34	0.12	-0.27	-0.17
45-54	0.24	0.02	-0.17
Qualifications			
University	0.43	-0.19	-0.34
Vocational	0.23	0.09	0.00
Day of the week			
Saturday	1.65 ***	0.59	-0.45
Sunday	1.53 ***	0.89 **	-0.57
R square	.139	.139	.075

* P-value<0.05 ** P-value<0.01 *** P-value<0.001 N=1690
Source: Author's calculations of ABS TUS 1997.

mothers 37 minutes more sleep than on weekdays. Fathers, but not mothers in either group, also enjoy extra sleep on Saturdays (25 minutes).

Personal Care

There is a considerable difference in the average amount of time working and non-working women spend in personal care activities such as eating, drinking, bathing, grooming and dressing (see Table 5.5). This is another activity in which working women average substantially less daily time than non-working women, meaning it could be redirected by working mothers to time in care of their children. Working mothers in the reference category spend, on average, just under two hours a day

Table 5.4   Coefficients of hours a day spent sleeping

	Sleep					
	Father		Mother			
			Employed		Not Employed	
Variable						
Constant	8.77	***	8.37	***	8.36	***
Type of non-parental care						
Mixed	-0.41		-0.28		0.15	
Formal only	0.00		-0.10		0.23	
Informal only	0.00		-0.14		0.17	
Duration non-parental care (hours a week)	-0.00		0.01	**	-0.00	
Market work (hours a week)	-0.00		N/A		N/A	
Spouse's market work (hours a week)	0.00		0.00		0.00	
Household income	-0.00	***	0.00		0.00	
Single parent	N/A		0.30		0.05	
Child >2 years	0.03		-0.00		0.24	
Number of children						
Two	0.00		-0.18		-0.18	
Three or more	-0.13		-0.25		-0.41	
Disabled person in household	-0.20		-0.17		-0.08	
Age						
25-34	-0.17		0.18		0.07	
45-54	0.16		-0.02		-0.01	
Qualifications						
University	-0.08		-0.55	**	-0.47	
Vocational	0.03		-0.29		-0.39	
Day of the week						
Saturday	0.40	*	0.29		0.37	
Sunday	1.10	***	0.70	***	0.62	**
R square	.098		.111		.068	

* P-value<0.05 ** P-value<0.01 *** P-value<0.001 N=1690
Source: Author's calculations of ABS TUS 1997.

in personal care activities whereas non-working women in the reference category average just over three hours a day. So on average working mothers squeeze an hour a day personal care time, which could be devoted to child care. Fathers do not sacrifice their personal care time to the same degree. They spend nearly two hours and twenty minutes a day in personal care which, though 48 minutes less than non-working mothers, is 25 minutes more than working mothers have.

The use of non-parental care does not predict that working mothers will be freed up to increase their time in their own personal care. Non-working mothers, in contrast, do gain personal care time from the use of extra-household child care. For every hour a non-working mother uses day care for her child, she adds 0.03 of an

Table 5.5   Coefficients of hours a day spent in personal care

	Personal care					
	Father		Mother			
			Employed		Not Employed	
Variable						
Constant	2.38	***	1.97	***	3.08	***
Type of non-parental care						
Mixed	-0.39		-0.00		-0.07	
Formal only	-0.22		-0.06		-0.33	
Informal only	-0.25		-0.06		-0.27	
Duration non-parental care (hours a week)	0.00		0.00		0.03	*
Market work (hours a week)	-0.00		N/A		N/A	
Spouse's market work (hours a week)	-0.00		0.00		0.00	
Single parent	N/A		-0.17		0.23	
Child >2 years	0.01		0.00		-0.00	
Number of Children						
Two	-0.00		-0.20		-0.55	**
Three or more	-0.04		-0.33	*	-0.67	***
Disabled person in household	-0.05		0.00		0.25	
Age						
25-34	0.01		0.13		-0.35	*
45-54	0.33		0.10		0.02	
Qualifications						
University	0.04		0.12		-0.32	
Vocational	-0.12		0.16		-0.30	
Household income	0.00		0.00		0.00	
Day of the week						
Saturday	0.20		0.38	*	0.33	
Sunday	0.43	**	0.31	*	-0.18	
R square	.036		.051		.091	

* P-value<0.05 ** P-value<0.01 *** P-value<0.001 N=1690
Source: Author's calculations of ABS TUS 1997.

hour to her personal care time. This would mean an increase of 23 minutes a day for the average non-parental care (for non-working mothers who use care) duration of 13 hours a week. There is no difference in the time non-working mothers spend in personal care on the weekends than during the week. In contrast, both fathers and working mothers make up the deficit in their daily personal care time at weekends by spending, for fathers, 24 minutes longer on Sundays, and for working mothers, 22 minutes more on a Saturday, and 18 minutes more on a Sunday.

Child-free Recreation

Table 5.6 gives a graphic indication of the time pressure on working mothers. Working mothers in the reference category appear to get no child-free recreation at all. The constant term predicted by the model for working mothers in the base category is negative. There is a substantial difference between the situation of working mothers and that of either fathers or non-working mothers. The average child-free recreation time of fathers with children under 5 years old is an hour and 12 minutes a day. The average for equivalent non-working mothers is 24 minutes a day.

None of the independent variables, including child care use, is associated with an increase in child-free leisure time for working mothers. In contrast, using non-

Table 5.6   Coefficients of hours a day spent in child-free recreation

			Child-Free Recreation			
	Father		Mother			
			Employed		Not Employed	
Variable						
Constant	1.21	***	-0.19		0.38	**
Type of non-parental care						
Mixed	-0.10		0.00		0.64	**
Formal only	-0.01		0.17		0.26	*
Informal only	-0.20		0.28		-0.12	
Duration non-parental care (hours a week)	0.01	*	-0.00		-0.00	
Market work (hours a week)	-0.00		N/A		N/A	
Spouse's market work (hours a week)	-0.00		-0.00		-0.00	
Single parent	N/A		0.27		0.00	
Child >2 years	0.20		0.18	**	0.12	
Number of Children						
Two	-0.36	**	-0.00		-0.17	
Three or more	-0.37	**	0.00		-0.33	**
Disabled person in household	-0.14		0.00		-0.00	
Age						
25-34	-0.07		0.14		-0.21	**
45-54	0.31		-0.10		0.00	
Qualifications						
University	-0.58	***	0.01		-0.01	
Vocational	-0.39	***	0.05		0.03	
Household income	0.00		0.00		0.00	
Day of the week						
Saturday	0.37	*	0.05		0.12	
Sunday	0.18		0.06		-0.15	
R square	.063		.052		.095	

* P-value<0.05 ** P-value<0.01 *** P-value<0.001   N=1690
Source: Author's calculations of ABS TUS 1997.

parental child care does increase fathers' child-free leisure time. The gain in child-free leisure for a father of a child who spends 20 hours a week in non-parental care is 12 minutes a day. This suggests that when working couples do not use day care, the fathers are to some degree participating in child care and losing some child-free leisure time.

Non-working mothers also are predicted to gain child-free leisure from the use of extra-household child care. The use of mixed care and formal care predicts an increase in non-working mothers' child-free leisure of 36 minutes and 18 minutes a day respectively. Also, there is a very small but significant effect on child-free leisure of non-working women with household income. The model predicts that at a weekly income of $1,000 this amounts to an extra 20 minutes a day. No similar effect is found for working mothers. No mothers gain child-free leisure on the weekends. Fathers, in contrast, average 24 minutes more child-free leisure on a Saturday than on a weekday.

Taken together, the results of these analyses indicate that in order to preserve time with their children, working mothers average less time in housework, personal care and child-free leisure time than other parents. The use of non-parental child care does not assist working mothers to find more time in these activities than working mothers who use no child care at all. Using non-parental child care gives non-working mothers (if they can afford it) more daily time for personal care and for child-free recreation, but does not confer these opportunities upon working mothers. This implies that non-working mothers use child care to reschedule daily activities, but working women just give up the time, in order to direct it to either paid work or child care. Some of the time that working mothers lose in housework and personal care on a daily basis is being made up on weekends, which presumably further restricts time in leisure.

Time Shifting

The time squeezed by working mothers from the activities discussed above gives a partial answer to how working mothers find time to spend with children. However, it does not fully account for the gap between working hours, child care use and the time mothers spend with children. I now investigate whether in addition to daily non-work and non-child care activities being reduced, also maternal child care is being rescheduled around work commitments. In other words, are mothers who both work and use child care shifting the time they spend caring for their own children to earlier or later in the day?

Figure 5.1 shows the percentage of households performing active child care between 6.30 a.m. and 8.00 a.m. The black line represents households in which the mothers work full-time (35 or more hours a week). The dotted line represents households in which mothers do not participate in the paid work force. Until 8.00 a.m., the average participation rate in active child care is higher in households in which the mother is working full-time than in households in which the mother is not working. This shows that families with working

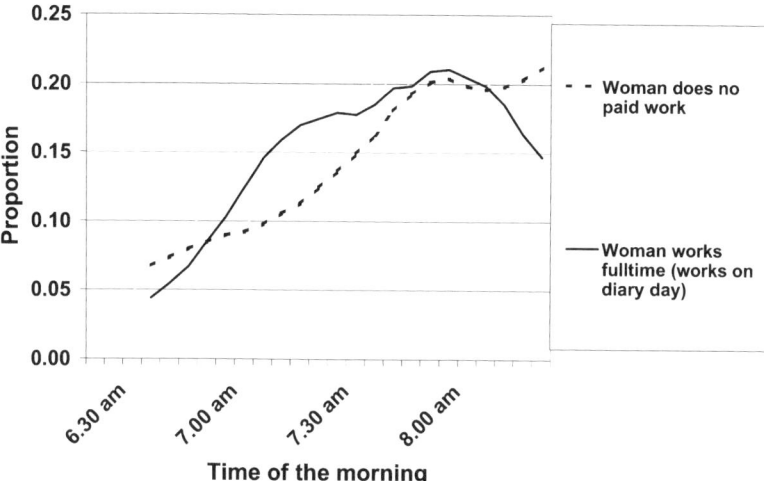

Figure 5.1 **Proportion of households participating in active child care by woman's workforce status (morning)**

Source: Author's calculations of ABS TUS 1997.

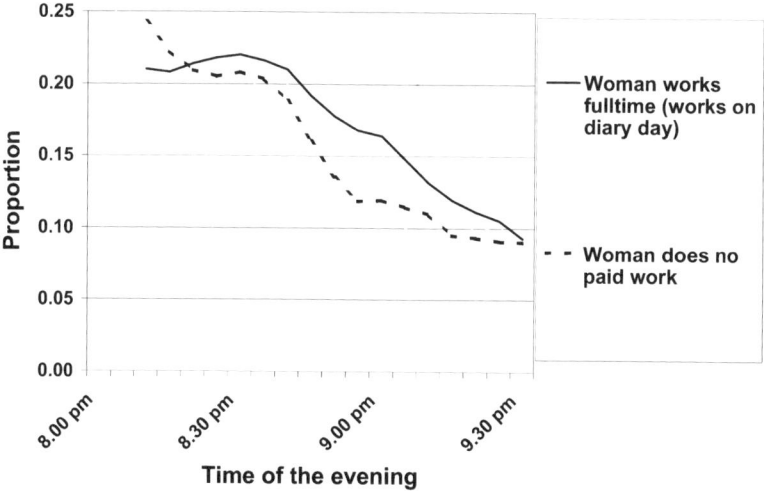

Figure 5.2 **Proportion of households participating in active child care by woman's workforce status (evening)**

Source: Author's calculations of ABS TUS 1997.

mothers begin their days earlier, and taper off their child care activity earlier in the morning than households with non-working mothers.

The same is true at the other end of the day (see Figure 5.2). Between 8.30 p.m. and 9.30 p.m., households with working mothers have a higher average participation rate in active child care tasks than households with non-working mothers. Parents in working mother households are more likely to be involved in active child care tasks after 8.30 p.m. than other families. This implies that children in some of these families are going to bed later than children in non-working mother households. It should be remembered that these data represent families with children under 5 years old. The phenomenon of time shifting may be more pronounced in families with older children.

Discussion and Conclusion

This chapter has canvassed research showing that non-parental child care does not completely replace mothers' time with their own children. Women do not put children into care and then lower their own caring time by an equivalent amount, but instead, substantially make up for time that children are with substitute carers. Quality time, in particular, is preserved or protected (Nock and Kingston, 1988; Bryant and Zick, 1996b; Sandberg and Hofferth, 2001; Booth et al., 2002; Bittman et al., 2004).

Given this, it is hardly surprising that working mothers of young children have been found to be the most time-pressured of all demographic groups (Craig et al., 2006). By looking closely at how they allocate time, this analysis gives ample reason why they would be constantly rushed and exhausted. In comparison with non-working mothers, employed mothers squeeze time from non-work and non-child care activities. They reschedule child care time to earlier or later in the day. Households with mothers employed full-time are likely to begin child care activities earlier in the morning, and end them later at night, than households with non-working mothers. These figures suggest a lived reality in which mothers get up early to shower and dress for work and undertake preparation for the day, organising the morning routines and helping their children to put on their clothes and eat their cereal, and shepherding them to the car to be dropped at day care in time for mum to get to work. They conjure a picture of women rushing from work to pick up their children from day care, cooking and bathing and feeding and talking to and playing with and reading to their children, and cuddling them to sleep at 9 o'clock at night, before dropping exhausted into bed themselves and beginning it all again the next day.

On average, employed mothers spend less time in housework and shopping than non-employed mothers. Using child care does not predict an increase in this time. These results fit with an image of a working woman stopping at the supermarket to buy food to cook for dinner after picking up her children from day care on the way home from work, and then facing the evening tasks of feeding, bathing and putting children to bed. Employed mothers of young children spend even less time in personal care than other parents do. An illustration of this would be working mothers of young children who complain they cannot remember the last time they coloured their hair or had a shower without sharing the cubicle with a toddler. Employed mothers do reschedule some of these activities. Some of the lost time in unpaid work and personal care is recouped on the weekends. The use of non-parental child care is associated with a slight increase in working mothers' sleep time. Employed mothers of pre-schoolers get almost no child-free recreation, and the use of non-parental care on a workday does not predict any increase. It seems that working mothers spend any leisure time they have with their children also present. On Saturdays fathers manage to find a little more child-free leisure time, but mothers do not spend leisure time away from their children even at weekends.

These results are a testament to the high value mothers place on spending time with their own children. They also raise the possibility that working mothers who use non-parental child care are more willing to contemplate adverse outcomes to themselves than to their children. Mothers may be avoiding an unacceptable trade-off between time in paid work and time in care of their own children, but if much of the time they preserve for child care is found by squeezing their own time in recuperative activities, it may be at some risk to their own welfare. Alternatively, it may be that women are simply unable to avoid spending this time with children. That is, if women wish to do paid work, they cannot get out of an equivalent amount of care time, but must add it to their workload, perhaps because they cannot persuade their partners to further assist. In the next chapter, I address the question of whether women with greater domestic bargaining power manage to achieve a different balance of paid and unpaid work or greater gender equity in child care time than other women.

Chapter 6

Earning Capacity versus
Caring Capacity[1]

A central pillar of the emancipation of women over the last 50 years has been their greater economic independence as a result of participation in the labour market. Analysis in the previous chapter showed, however, that even when mothers engage in market work, or use non-parental child care, they do not substantially reduce their time with children. In this chapter I investigate whether this finding is similar across variation in parental education level.

Background

Economic theories, especially the human capital variants, suggest that when women have greater choices in the market they will reduce their time commitment to the home. People with higher human capital will allocate more time to paid than to unpaid labour because the opportunity costs of foregoing wages are higher (Becker, 1981; Becker, 1985; Becker, 1991). However, the historical masculinisation of women's work patterns has been stronger than any masculinisation of female care responsibilities. In this chapter I investigate whether parents with the most potential or actual economic freedom, those with high education and therefore high earning capacity, allocate their time to children and to the market differently from the ways other parents do. I begin with a theoretical overview of economic theories that relate to labour market decisions or touch on time allocation to children, of historical developments in domestic arrangements and ideas about childrearing, and of the effects that education is thought to have on gender equity. I identify the conflicting imperatives that may affect parents who have invested in their own human capital but may also subscribe to an ideal of high investments in their children's human capital. It may be that the parents who (human capital theorists would suggest) experience the strongest pull to the market, those with high education and earning capacity, are also those most pushed towards fostering the development of their children in accordance with current views on attentive childrearing. I investigate at both a household level and separately for men and women, whether education and earning capacity have implications for time with children and how child care is divided between partners.

---

1   Parts of this chapter appear in Craig, Lyn (2006) 'Parental education, time in paid work and time with children: An Australian time-diary analysis' *British Journal of Sociology* 57 (4): 553–575 © 2006 London School of Economics and Political Science.

The Pull to the Market

Labour supply

Economic theories conceptualise the individual as a rational agent, making labour participation choices that involve selecting the point at which the utility derived from market goods and non-market time is optimal, given tastes, available options, and budget constraints (Killingsworth and Heckman, 1986; Blau et al., 1998). The labour supply decision is theorised as a rational calculation of how best to spend time to obtain the most benefit. Neoclassical economics assumes that people work for money. The time spent in employment is allocated to the market solely in order to obtain money to spend on goods or services. Conversely, non-market time is regarded as devoted to consumption, and is classified as leisure. Rational agents allocate their time to work or leisure in the ratio that would give them the highest utility, with the wage rate the major criterion for making this judgement (Mincer, 1962; Becker, 1965). The trade-off is between work at a given wage rate and the leisure or consumption opportunities foregone (Killingsworth and Heckman, 1986; Blau et al., 1998).

Female labour supply

The theoretical stance of economics is gender neutral and individualistic. However, labour economics does distinguish between male and female labour supply patterns. Whereas men are regarded as single individuals making rational choices to maximise personal utility, women's choices are regarded as more likely to reflect family circumstance. Theories of labour economics and labour supply suggest that, due to their domestic responsibilities, women are less firmly attached to the labour force than men (Mincer, 1962; Mincer and Polachek, 1974). They acknowledge that women have a disproportionate responsibility for unpaid work so the conceptual division of time into the two categories of market work and non-market consumption is adequate to explain male time allocation, but not female (Blau et al., 1998). Whereas men allocate time between work and leisure, women allocate time between work, leisure, and unpaid work (Mincer and Polachek, 1974; Killingsworth and Heckman, 1986).

Because women have more alternative uses for their time, the female labour supply is more elastic than the male labour supply (Mincer, 1962; Killingsworth and Heckman, 1986). This is with regard both to whether they will be in the work force at all (participation rate), and to the 'hours of work decision' of how much time to allocate to paid work (Blau et al., 1998). Labour supply theory suggests that the reservation wage, the point at which a woman is just as likely to work as not, is equal to the value she places on her time at home (Mincer, 1962). This point will vary with her circumstances, because female labour supply is much more sensitive to variation in family composition than male labour supply (Killingsworth and Heckman, 1986; Blau et al., 1998).

Human capital

The wage rate an individual can command is influenced by their stock of human capital. The idea of human capital, originally applied to measure the value of education and expertise at a national level, is at a personal level the attributes of the individual's education, training and on-the-job experience that influence their earning capacity in the job market (Mincer and Polachek, 1974; Becker, 1994). The cost of acquiring human capital involves not only direct expense, such as paying for university and training courses, but also the sacrifice of earnings in the short term in order to accumulate qualifications or experience that will command higher earnings in the long term. The assumption is that the greater the investment a person has made in their marketable skills, the higher the wage rate they will command and the stronger will be their attachment to the paid work force (Becker, 1965; Mincer and Polachek, 1974; Garfinkel and Haveman, 1977; Becker, 1994).

Human capital theory suggests that the increase in female education since the mid-twentieth century should lead to greater female labour force attachment (Mincer and Polachek, 1974). Because education leads to higher earnings, it increases the discrepancy between the value of employed time and the value of home time. Therefore, because the opportunity costs of staying at home are higher the higher the amount of earnings foregone, those with higher levels of education are more likely to be in the work force (Blau et al., 1998). The theory predicts that, as women generally become more educated, female labour supply will grow (Mincer and Polachek, 1974; Killingsworth and Heckman, 1986; Becker, 1994).

New home economics (discussed in greater detail in Chapter 1) sees education as being in direct conflict with the mothering role, arguing that educated women are likely to have fewer children because the opportunity costs of leaving the work force are higher for the more highly educated (Becker, 1981; Becker, 1991). It predicts that declining fertility will result when women make significant investments in their own human capital. Because her time is more valuable in the market, the opportunity cost of an educated woman caring for children is higher than that of a woman with less education. Therefore, the theory argues, an educated woman will have fewer children in order to direct more of her time to paid work (Becker, 1991). The view is supported by demographic trends. Highly educated women *are* more likely than others to remain childless (ABS, 2005). Highly educated women who do become mothers are likely to be older at the time they first give birth, and to have fewer children in total than their less well-educated counterparts (Andorka, 1978; Astone et al. 1999; Caldwell, 1999; Barnes, 2001). This means that at a national level, increasing female educational opportunity is associated with lower birth rates (McDonald, 2001; McDonald, 2004a). However, it is not clear if human capital theory is as helpful in predicting how parental investment in human capital will affect time inputs to children once born.

Earnings capacity

There is a distinction between the maximum value of earnings that could be generated from the stock of human capital, at both a national and a household level, and what is in fact generated (Garfinkel and Haveman, 1977; Haveman et al., 2002). This has led economists to suggest that, rather than focusing on actual income or consumption patterns, what should be calculated is earnings capacity. Central to the method is the application of the two core human capital variables, years of experience and level of education. Household earnings capacity is what would be earned if all working-age household members worked on a full-time full-year basis consistent with their age, education and other characteristics, adjusting for child care costs (Haveman and Bershadker, 1998). Households vary in their preferences for leisure, and in their amount of domestic production. Earnings capacity measures reflect a family's potential to achieve independence rather than whether they actually do so (Garfinkel and Haveman, 1977). The implication is that not using their full earnings potential is an active choice that should not require social compensation. Applied to households with children, this implies that the only limitations upon all the adults reaching their full earnings potential are child care costs and personal choice.

Bargaining theories

The economic theories above do not deal with how decisions about labour supply are negotiated within the family. Bargaining theories, briefly discussed in Chapter 1, do address how marital partners distribute their time between paid and unpaid work. These theories are derived from economic game theory (Manser and Brown, 1980; McElroy and Horney, 1981; Lundberg and Pollack, 1996), and sociological exchange theory (Scanzoni, 1979; Molm and Cook, 1995). Bargaining theories conceptualise intra-household decision-making on work allocation as resulting from bargaining between partners, on the basis of their relative resources (Manser and Brown, 1980; McElroy and Horney, 1981; Lundberg and Pollack, 1993; Brines, 1994; Lundberg and Pollack, 1996). The idea is that, as marital partners become more equal in resources, they will divide domestic tasks, including child care, more equally. An individual's bargaining power is increased by his or her threat point(s). If they fail to reach agreement, both husband and wife receive the utilities associated with a default outcome. This is the threat point (Manser and Brown, 1980; McElroy and Horney, 1981; Lundberg and Pollack, 1996). External threat points are those at which a partner would see the benefits of leaving the marriage outweighing those of remaining and would end the marriage (Manser and Brown, 1980; McElroy and Horney, 1981). Internal threat points are not divorce but non-cooperation within a continuing marriage. They relate to what one spouse can withhold from the other without leaving the marriage. An increase in threat point results in family arrangements that more closely reflect the preferences of that spouse (Lundberg and Pollack, 1996). It is suggested that traditional gender roles are the default or fallback position if new agreements cannot be made (Lundberg and Pollack, 1993; Lundberg and Pollack, 1996; Widmalm, 1998).

The typical gender division of labour disadvantages women in bargaining within marriage (England and Folbre, 2002). A labour market in which women are systematically paid less than men leads to domestic inequality because wives earn less than husbands and therefore have less power at home, and conversely, if women do specialise in home duties, their skills are less readily transferred to another arena and this also reduces their bargaining power (England and Kilbourne, 1990). Bargaining theories admit scope for acknowledgement of public interest in children. When the end of a marriage is contemplated the social policy environment is an important constitutive part of the threat point, which comes into sharpest focus if the household includes children. Overwhelmingly, women retain custody of children following marital breakdown. A policy environment that regards them as a private responsibility is less likely to assist sole parent households in the care of their children than one that acknowledges the social value of children (Millar and Rowlingson, 2001; Daly, 2002). So the threat points depend on the individual resources of each spouse and come from a range of sources, including legal and institutional factors – such as family and property law – and social entitlements that would affect economic well-being following divorce, as well as personal attributes including education level, income and earning capacity, and the degree to which they can turn them to personal advantage (McElroy, 1990; Lundberg and Pollack, 1993; Folbre, 1997).

This last aspect of the theory – *the degree to which partners can turn their individual resources to personal advantage* – points up a limitation of viewing time investment in children from an economic perspective. A calculation of potential earnings that could arise from a parent's education or job experience does not necessarily equate with their ability to reach this potential. The wish or necessity to care for children may inhibit mothers particularly from capitalising on any advantage they could potentially have in the market place as a result of their own education and/ or earning capacity. Well-educated mothers may have a theoretical earnings capacity that they are unable to turn to personal advantage. While the individualistic approach of economics would attribute this to personal choice, it may arise for more complex reasons, including workplace and institutional constraints, and also norms, gender attitudes and cultural factors. Not least, personal decisions about time with children will be influenced by how the responsibilities of parenthood are socially defined. I now trace the development of the latter over the last century.

The Push to the Home

During the period in which women have become more highly educated, and more able to support themselves through paid employment, there have been enormous change in the social expectations of responsible parenting (de Mause, 1974; Shorter, 1977; Donzelot, 1979). Theories of child development and psychology suggest that a high level of parental education and expertise in child rearing practices is desirable (Oakley, 1979; Cowan, 1983; Reiger, 1985; Ehrenreich and English, 1989; Mein Smith, 1997). The specific advice on best parental practice has varied, but it has consistently included advocacy of a great deal of personal input from informed parents to children, and argued that maternal bonding, attentive parenting and high time inputs are necessary for optimal educational and social outcomes for children (Bowlby, 1972; Bowlby, 1973; Leach, 1977; Spock, 1998; Belsky, 2001; Leach et al., 2005).

The creation of full-time motherhood

Following the industrial revolution, women and children, who had been active economic agents, were progressively excluded from the new paid workforce. As briefly canvassed in Chapter 1, by the end of the nineteenth century, industrial organisation had produced the male breadwinner family model. Home production, as economically important as paid employment but no longer so recognised as such, had been forced into a newly defined unpaid sector (Cowan, 1983; Eisenstein 1979; Folbre, 1991). As a by-product, raising children was also forced into the unpaid sector. What had been achieved as an integral part of economic and family life became a discrete non-economic undertaking.

Technological changes were associated with the transfer of male jobs from the home to the factory and rendered child domestic labour negligible. In contrast, technological changes to female jobs were not associated with a move outside the home. The technology of female work was transformed, but not the location (Cowan, 1976; Cowan, 1983). What changes to domestic technology did mean was that running the home no longer required the simultaneous labour of several people. Servants were no longer needed (or readily available). This meant a major change in the duties of middle-class women for whom running a household had been a matter of management rather than hands-on labour. Conversely, for the working class, the rise in living standards resulting from larger housing increased, rather than reduced, domestic demands (Bourke, 1993). Class distinctions in women's lives became less marked. In a sense, all housewives became proletarianised, in that they all performed similar domestic labour. One person per household could now manage female tasks, but for that one person they were more isolating, and arguably more time intensive, than they had previously been (Cowan, 1976; Cowan, 1983).

Not coincidentally, household management became a matter of professional concern. The domestic economy movement burgeoned, constructing a new model of the efficient housewife undertaking scientific management of the home (Reiger, 1985). A major campaign of expert information was directed at the organisation of work in the household and at appropriate ways of family living (Cowan, 1976; Cowan, 1983; Reiger, 1985). While some household labour was reduced with mechanisation (laundry, for example), new standards of cleanliness and nutrition meant that in many cases new tasks replaced the old (Cowan, 1983). Furthermore, housework became regarded as no longer simple work, but as the means through which a good wife and mother expressed her love (Oakley, 1981). Information dissemination such as advertising of products necessary for reaching the new standards of domestic management assisted the professional discourse in establishing new behavioural norms (Cowan, 1976; Cowan, 1983; Reiger, 1985; Gilding, 1991).

The new outlook also pertained to child rearing. The increased domestic demands associated with improved living standards were paralleled by a rise in expectations of the job of parenting. Parenting in the late nineteenth and early twentieth centuries was a matter of economic interdependence and moral guidance. During the twentieth century, children were reconstituted as a new object of concerned attention (de Mause, 1974; Shorter, 1977; Donzelot, 1979), with motherhood defined as a rationally controlled learned activity (Reiger, 1985). Child welfare became a matter of professional interest and motherhood a set of skills that had to be taught by experts (Cowan, 1976; Donzelot, 1979; Oakley, 1979; Cowan, 1983; Reiger, 1985). Originally intended to reduce childhood illness and death, the teaching of mothercraft was associated with an enormous increase in the work required to be good mother. Advice from experts proliferated: how to be an informed consumer, how to provide good nutrition, how to avoid disease through new standards of cleanliness such as sterilising nappies and bottles (Oakley, 1979; Reiger, 1985).

The prevalent view of child rearing became that 'children require constant attention from well informed persons' (Reiger, 1985, p.35). The concern, originally for children's physical well being, came to encompass psychological considerations. Further, it was thought paramount that the person delivering care to children was their own mother (Bowlby, 1953). Responsibility for the psychological and emotional development of children was laid at their mother's door (Reiger, 1985). This further entrenched the similarity of women's responsibilities across social class (Cowan, 1976; Cowan, 1983). The idea that good motherhood required constant presence meant both that middle-class women could no longer delegate and, conversely, that working-class women were encouraged to stay at home to care for their children rather than to take employment. Mothering, rhetorically natural and instinctive, was subjected to new pressure that implicitly made it more difficult, anxiety-prone and complicated (Ehrenreich and English, 1989; Mein Smith, 1997). To equip women for this formidable task, they required education, and it was for this reason that the education of women was first introduced.

Female education was originally intended to school women in domestic science, either to be informed and competent full-time mothers, or to take up such jobs as infant welfare nurses disseminating the prevailing expert view (Ehrenreich and English, 1989). The recurrent themes were maternal responsibility for children's problems, and a consequent need for women to be educated about child development (Reiger, 1985). Children's requirements were by then seen to include special play opportunities, toys and equipment, for which houses with backyards and a separate room for each child were desirable. This meant both more domestic labour and more financial outlay. The child-rearing job description expanded to include supervising the development of social skills, facilitating friendships, and providing a secure emotional environment for which a happy household, preferably including a stable marriage, was regarded as an essential condition.

Home inputs to human capital

Of increasing importance was parental facilitation of children's education. During the twentieth century, children's economic contribution to families disappeared almost completely (de Mause, 1974; Caldwell, 1982; Zelizer, 1985). Following the introduction of compulsory schooling, the major task of childhood became the acquisition of education. School hours were predicated on the assumption that there would be a mother able to care for children before 9.00 a.m. and after 3.00 p.m. Children spend years being educated, to the point where it is now common to continue tertiary studies well into one's 20s. Parental support of children thus extends over a longer period, and increasingly includes the purchase of educational products such as music lessons, tutoring, organised sport classes and private schooling. The expectations of good parenting now include encouragement of school participation, including fostering study habits and monitoring of homework. Regulation has followed: in the UK parents are required to sign contracts agreeing to ensure their children attend school and complete their homework (Williams, 2006).

These inputs are clearly an important contribution to a person's human capital acquisition. England and Folbre (1997) argue that mainstream economic theory ignores such family involvement by narrowly classifying human capital as tertiary qualifications and experience (see for example Mincer and Polachek, 1974; Becker, 1975). They suggest that this means the expectations and pressures upon the family to perform this function are both powerful and unacknowledged. Conventional human capital theory ignores private sphere contribution to human capital development, despite family inputs being of major importance (England and Folbre, 1997).

Nor does conventional human capital theory look at how the acquisition of human capital may impact upon values and therefore upon parental inputs to child care. It is possible that educated parents would be among those most receptive to ideas disseminated by professionals on child development, psychology and education, and most likely to incorporate them into their child-rearing practices. Parents who have invested most highly in their own human capital may be those who most want to build up their children's. If so, the pull effect towards capitalising on earning potential in the market may be counteracted by a push towards increasing time inputs to child care. Parents with high human capital may spend either more time doing child care than other parents, or more time in the child care activities, such as talking, reading, and playing that arguably most assist the human capital acquisition of their children (Brooks-Gunn et al., 1993).

There has, however, been little empirical investigation of how all the expert advice on childrearing was received. It is easier to trace the information stream than to establish how it was adopted in practice. The extent to which these ideas were taken up is not available from the prescriptive literature. However, it cannot be assumed that all parents followed all the advice. Mothers were not simply passive recipients of expertise, but on issues such as the use of dummies and the amount of appropriate cuddling, chose advice from various sources including infant welfare nurses, mothers, and friends (Reiger, 1985). Further complicating the picture is that the expert advice was by no means consistent, but varied enormously as intellectual fashions changed (Bittman and Pixley, 1997).

It also contained internal contradictions. Psychological theories of child development were predicated on the idea that each individual should maximise their personal potential. This did not mesh with the sacrifice required of mothers unless females were to be excluded from these (supposedly universal) theories. The contradiction became glaringly apparent as girls brought up to pursue opportunities for eduction and independence reached childbearing age. Ehrenreich and English (1989) argue that, by the 1960s, the contradiction between the child welfare experts' promotion of maternal self-sacrifice and the culture of the individual and self-gratification meant that full-time motherhood could only be explained as masochism. To be a housewife was in some sense insane. Women who committed themselves to maintaining the home full-time were viewed as making a choice, but because the choice was so clearly not in their own best interests, they were therefore aberrant martyrs (Ehrenreich and English, 1989; Hakim, 2000).

As education became more widely accessed, expert advocacy of full-time mothering became increasingly in conflict with women's other aspirations (Mitchell, 1971; Oakley, 1979; Ehrenreich and English, 1989). Children were no longer the only source of female fulfilment but an obstacle to economic independence and self-actualisation. There were practical contradictions too. Human capital is embodied in the individual, and is not transferable between partners. Economic dependency has become increasingly risky for women as social changes, such as rising marital dissolution rates, mean that women who specialise in housework are making a marriage specific investment that places them at a potential disadvantage. If a woman has experience only in skills specific to a marriage, she is not able to transfer that knowledge to another situation should the marriage end (England, 1993; Ferber and Nelson, 1993; Folbre, 1994b; Nelson, 1996; Bittman and Pixley, 1997). This is particularly so in the case of time investment in children (England and Kilbourne, 1990; Blau et al., 1998).

The contradiction between mothers' and children's welfare became evident. This has never been satisfactorily addressed and is one of the strands that underpin the current mothering predicament. We have developed a culture in which the benefits of reproduction to the whole of society are rarely articulated. The implied pathology of the self-sacrifice involved in motherhood means the desire to have children has become hard to give reasons for, except as a personal choice for the anticipated 'process benefits' or pleasure of caring for them (Juster and Stafford, 1991). In accordance with new home economics' conceptualisation that those who had children have revealed a preference that implied an acceptance of the associated costs (Becker, 1981), childrearing is more each woman's individual problem, than a public priority or responsibility (Reiger, 1985; Folbre, 1994b). Theory has failed to articulate a response to the dilemma of how educated women who may have high earning capacity can both reach this potential and at the same time meet the needs of their children (Riley, 1983). How are mothers both to develop fully their children's human capital and to recoup their investment in their own?

Involving fathers

The usual response is to seek to increase male input, and to try to share the care. As discussed in Chapter 4, good fathering is increasingly being defined as necessarily active and involved. There are some signs that this agenda is most advanced among men with higher education. Research has found that higher parental education is associated with decreased time in unpaid work for women and increased time in unpaid work for men (Goldscheider and Waite, 1991; Bryant and Zick, 1996b; Bianchi and Robinson, 1997; Robinson and Godbey, 1997). However, the effects are slight, and it is not clear whether the effect pertains to both housework and child care. It may imply weak support for bargaining theory or may result from a more indirect effect, in that education may be instrumental in altering ideas and values. While the social acceptance of gender equity is far stronger than is reflected in behaviour (Hochschild and Machung, 1989; Bittman and Pixley, 1997; Dempsey, 1997), for both men and women, education level is positively associated with support for gender equality in paid and unpaid labour (Bittman and Pixley, 1997; Dempsey, 2001; Baxter et al., 2005).

In summary, whereas economic theory sees human capital investment as predictive of greater workforce participation, and education as a resource that could be used to bargain down one's domestic labour burden and/or improve one's earning capacity, social history suggests that the effects of education may be more ambiguous. Education can reflect or influence attitudes, and also, for women, was historically intertwined with motherhood (Cowan, 1983; Reiger, 1985; Ehrenreich and English, 1989; Mein Smith, 1997). At the same time as educational and work opportunities for women have increased, so have social expectations of the parenting role (de Mause, 1974; Shorter, 1977; Oakley, 1979; Cowan, 1983; Reiger, 1985; Ehrenreich and English, 1989; Mein Smith, 1997). It is therefore possible that higher levels of education, of both men and women, would be linked with receptivity to ideas on involved, hands-on childrearing. Also, parents who have invested in their own human capital may be the ones most concerned with their children's acquisition of human capital (Blau et al., 2000). This perspective would predict that higher education would be associated with higher parental time devoted to child care, particularly in interactive care activities. Higher human capital investment may be associated not only with a pull to the market, but also a push to the home.

Research Focus and Method

In this chapter, I provide a snapshot of daily time-use that shows the effect of human capital investment on direct parental time inputs to child care, first at a household level, and then, to investigate gender equity, at an individual level. I use education variables as a proximate indication of earning potential. I do this because I am most interested in how parents actually allocate their time despite variations in what they *could* earn if they gave market time the highest priority.

Using the TUS and the regression model described in Chapter 2 and summarised in Table A1, I investigate four possible aspects of how, consistent with the literature, education may affect time in paid work and child care:

1. Investment in their own human capital attracts parents to the workforce and therefore has an indirect downward impact upon parental time inputs to children.
2. Higher education is associated with receptivity to expert opinion on the importance of involved and attentive childrearing and therefore has a direct upward effect on parental time inputs to children.
3. Higher education levels are associated with greater acceptance of the ideals of domestic gender equity so partners with higher levels of education will share (a) housework and (b) child care more equitably than other parents.
4. Parents with higher education levels put more time than other parents into the type of child care (interactive) that most fosters the human capital acquisition of their children.

The measures investigated are as described in Chapter 2: daily time spent in child care (done as primary activity only or as either a primary or a secondary activity), in housework and in paid employment (amounting to total workload) and daily time spent in the sub-categories of child care, 'interactive care', 'physical care', 'passive care' and 'travel and communication'. At the household level the child care variables represent primary activity only. At the individual level I present two sets of results, one counting primary activity and the other primary and/or secondary activity. As discussed in Chapter 2 and more fully in Chapter 4, the latter is a more accurate account of parental time allocation to children. This is particularly true of low intensity (passive) child care, but a substantial proportion of interactive care, and some physical care is also done as a secondary activity. Including secondary activity in the sub-categories of child care also gives a valuable comparison of how child care is distributed on gender lines. For the purpose of this chapter, it will show whether mothers and fathers with varying levels of education do more double activity than other parents, and if this is similar for each sex. To see if educational attainment has any effect on whether men are more likely to assume responsibility for the job of child care (as discussed in Chapter 4), I use as a dependent variable the proportion of all time in the company of children during which the parent is not in the company of their spouse (that is, the time in which one parent is in sole charge).

The constant terms in the household regression results below represent time spent in the dependent variable on a weekday in a household in which there is a child under 5, the husband works full-time, the wife works part-time (the modal category), in which no non-parental child care is used, both partners are aged 35–44, and in which there is no disabled household member and neither partner has post-school qualifications. The constant terms in the individual regression results represent time spent in the dependent variable on a weekday by a parent of one child under 5, who has no tertiary qualifications, uses no non-parental child care, is aged 35–44, and has no disabled household member. The model is run separately for men and for women. For men the default labour force status position is 'employed full-time'; for women it is 'employed part-time'. When paid work is the dependent variable, labour force status is not included as an independent variable.

Results

Parental education and time in work and care (couples jointly)

The higher the qualifications of the marital partners, the more time the couple devotes to children. Families in which one partner has university qualifications spend more than half an hour a day longer in primary child care activities than less qualified families. When both partners have a university qualification, the amount of time devoted to child care is nearly an hour a day longer than in families with no qualifications or with vocational qualifications.

The effect of education on child care time is even more pronounced when child care performed as a secondary activity is included in the measure, and extends further, to include families that have vocational qualifications. Couples in which both

Table 6.1   Coefficients of couples' joint hours a day in child care, housework and paid work by qualification levels

	Childcare (Primary)	*Childcare (Primary & Secondary)*	*Housework*	*Paid Work*
Constant term	4.40 ***	8.52 ***	4.86 ***	6.75 ***
One partner has vocational qualifications, other partner has no qualifications	0.15	0.67	0.54	0.22
Both partners have vocational qualifications	0.41	1.45 **	-0.33	0.02
One partner university educated, other partner has vocational qualifications, or no qualifications	0.51 *	1.53 **	-0.04	-0.64
Both partners have university qualifications	0.84 ***	2.48 ***	-0.55	0.00

* P-value<0.05 ** P-value<0.01 *** P-value<0.001
Source: Authors calculations of ABS TUS 1997. The figures in this table are drawn from Table A5.

partners have vocational qualifications do over an hour and a half more child care as either a primary or secondary activity than less qualified couples. The effect increases incrementally as qualification levels rise. One partner being university educated is associated with an increase in child care as either a primary or a secondary activity by over an hour and a half daily. Both partners being university educated is associated with nearly two and a half hours a day more devoted to child care activities than in families in which neither parent has tertiary qualifications (see Table 6.1).

The model predicts no impact upon the amount of domestic labour or paid work time performed by couples with variations in qualification levels. More highly educated couples with children neither spend more time overall in market work, nor do less housework than couples who have fewer qualifications. These results suggest that higher education is associated with a direct upward effect on parental time inputs to children, but that it does not affect household time devoted to other types of work either paid or unpaid.

Table 6.2 shows to which child care tasks educated households direct their extra household time investments in child care. In most cases, the extra child care time is allocated to physical care. All couples with joint qualifications at vocational level or above spend about half an hour more in physical care than less qualified couples. Couples in which both spouses are university educated also allocate more time to interactive care, spending 14 minutes a day more than the average of half an hour that unqualified parents give to this activity. No extra household time in travel and communication or in passive care is associated with variation in educational attainment.

Taken together, the results in Tables 6.1 and 6.2 show that households with higher education devote more time to child care than other households do. This implies that educated households do place a higher priority on direct parental inputs to children

Table 6.2   Coefficients of couples' joint hours a day in sub-categories of child care (as a primary activity) by household qualification levels

	Physical Care		Interactive Care		Travel/ Communication		Passive Care	
Constant term	2.02	***	1.09	***	0.3	*	0.81	*
One partner has vocational qualifications, other partner has no qualifications	0.05		0.08		0.00		0.01	
Both partners have vocational qualifications	0.34		0.13		-0.12		-0.00	
One partner university educated, other partner has no qualifications	0.40	*	0.11		-0.03		0.01	
One partner university educated, other partner has vocational qualifications	0.70	**	0.39	*	0.00		0.04	
Both partners have university qualifications	0.46	***	0.21	**	0.00		0.03	

* P-value<0.05 ** P-value<0.01 *** P-value<0.001
Source: Author's calculations of ABS TUS 1997. The figures in this table are drawn from Table A4.

than do less-educated households. While there is no definitive causality, this may show that the expert dialogue on child care is taken up differently in households according to their levels of educational attainment. That the major variation is in the amount of physical care is consistent with the idea that the message that parents themselves should perform the close contact work (which includes menial and repetitive tasks such as bathing, feeding and changing diapers) has been particularly well absorbed by households with educated parents.

The child care activities of talking, playing, reading, teaching and reprimanding (interactive care) have been identified as contributing most directly to the development of children's human capital, as discussed in Chapter 5 (Brooks-Gunn et al., 1993; Shonkoff and Phillips, 2000). It is couples in which both partners are university-educated that allocate most time to these activities. Put another way, it is parents who have invested most highly in their own human capital who contribute most time to the child care activities that will most assist the human capital acquisition of their children. The prediction that the higher one's education and earning potential, the higher the opportunity cost of child care, and therefore the lower the time allocation to it (which may have an indirect effect on child care time inputs by contributing to lower birth rates), cannot be extrapolated to predict the direct effect of education on child care time in families who do have children. This suggests that educational attainment is associated with a push to devote time to children.

This implies that the earning capacity approach, in which net earning capacity is measured as the income a family would generate were it to use its human capital to full potential, less child care costs (Haveman and Buron, 1993; Haveman et al., 2002), may be overlooking an important issue related to tastes, the least well-developed aspect of economic theory as it relates to work and family (Blau et al., 2000). The earning capacity approach suggests that time and expertise can be turned into money unless the person's tastes direct them otherwise. The implication is that people should not be compensated if they choose not to maximise their earnings potential, because they have revealed a preference for non-market activity. Adjusting for child care costs assumes that parents can choose to substitute others' care for their own if they want to. But recall that using non-parental care does not relieve parents of child care on an hour for hour basis. This may be because it is not possible to delegate all tasks, but also may be because parents want to maintain time in certain activities. That interactive care time is not lower when non-parental child care is used lends weight to this interpretation. Further, experts have argued that child care is not of equal value when performed by substitute carers rather than parents themselves (Cowan, 1983; Reiger, 1985; Ehrenreich and English, 1989; Mein Smith, 1997). This is a widely held view, despite the persistence of economic theory in ignoring it (Folbre and Nelson, 2000; England and Folbre, 2003). The results above suggest that earning capacity in the form of education may not be fully used because higher education is associated with other imperatives, such as investing human capital in the children, or in aspiring to give a high level of parental attention.

To summarise, education is associated with higher household input to physical child care, and at the highest education level, with an increase in the type of child care (interactive) most associated with human capital development. I now investigate how these inputs are distributed by gender.

Parental education and time in work and care: A gender comparison

Both men and women spend more time doing child care when they have higher qualifications, with the effect on female behaviour more pronounced than the effect on male behaviour (see Table 6.3).

The model predicts that all mothers who have educational qualifications will spend about 20 to 40 minutes a day more time doing child care as a primary activity than unqualified mothers. Broadly speaking, the higher her educational qualification,

Table 6.3   Coefficients of hours a day spent by fathers and mothers in child care, domestic labour and paid work by qualifications

	Childcare (Primary)		Childcare (Primary and Secondary)		Housework		Paid work	
	Father	Mother	Father	Mother	Father	Mother	Father	Mother
Constant term	0.86***	2.64***	1.40***	35.86***	1.08***	3.56***	5.10***	0.90***
Qualifications								
Vocational	0.02	0.02	0.69	0.25	-0.13	0.10	0.66	0.54
Skilled vocational	0.10	0.41*	0.28	0.92**	0.06	-0.36*	0.88**	0.38
University diploma	0.16	0.66***	0.62*	1.49***	0.06	-0.28	0.17	-0.24
Bachelor degree	0.37**	0.28**	1.24***	1.10***	0.02	-0.57**	0.37	0.82**
Postgraduate	0.37*	0.61*	1.22***	1.19*	-0.18	-0.55*	-0.28	1.16**

* P-value<0.05 ** P-value<0.01 *** P-value<0.001

Source: Author's calculations of ABS TUS 1997. The figures in this table are drawn from Tables A6 and A7.

the longer a mother devotes to child care as a primary activity. When secondary activity is included in the child care variable, the effect holds for mothers with skilled vocational qualifications or higher, and does not go up incrementally with qualification level.

For fathers, too, having more education is associated with more time doing child care. The effect occurs higher up the ladder of educational attainment than it does for mothers, kicking in for child care as a primary activity when fathers have a bachelor or postgraduate degree. In both cases, fathers are predicted to spend approximately 20 extra minutes a day in primary child care. All fathers with a university education spend longer doing child care if secondary activity is included in the calculations, with the extra time ranging from just over half an hour for those with a university diploma to about an hour and 20 minutes for those with a postgraduate degree.

The model predicts no variation in fathers' time in housework with educational attainment. This suggests that previous research showing that male time in unpaid work is positively associated with education (Bryant and Zick, 1996b; Bianchi and Robinson, 1997; Robinson and Godbey, 1997) may have had different results had child care and housework been disaggregated. In this study I find that while educated fathers are devoting more time to child care than other fathers, they are not devoting more time to other types of unpaid labour. In contrast, female housework time does vary with education level. Mothers with either a bachelor or a postgraduate degree spend about half an hour a day less time doing housework than other mothers. This suggests that educated mothers are redirecting into child care activities the time their less educated sisters spend in housework. It may also indicate that educated women will buy themselves out of domestic duties more readily than they will buy themselves out of child care.

Qualifications are not associated with any significant difference in the time fathers spend in paid employment, except for those with skilled vocational qualifications. It is not the case that the more educated a father is, the more time he allocates to paid work. In contrast, educational attainment *is* associated with an increased time allocation by mothers to paid work. With the exception of those with university diplomas, mothers with an educational attainment of skilled vocational qualifications or above allocate more time to market work than unqualified or vocationally unskilled mothers. The model predicts that time allocation to paid work will be higher the more educated a mother is. At its most pronounced, mothers who have postgraduate degrees average an extra hour and 10 minutes a day in paid employment.

The results indicate that, for mothers, educational attainment is associated with a higher allocation of time to market work, just as human capital theory would predict. At the same time, consistent with the idea that educated people would be most receptive to ideas of the importance of involved parenting to child development, both fathers' and mothers' child care time is positively associated with qualification level. So despite higher education being associated with more maternal time in market work, this does not translate to a lower time investment in children by educated mothers. The results suggest that higher education is associated both with higher direct time inputs to children, and with higher time inputs to paid work. Educated women appear to trade off paid work time against housework time, but not against child care time, which they add to their working day. Educated fathers spend more time doing child

Table 6.4   Coefficients of minutes a day spent by fathers and mothers in sub-categories of child care (primary and secondary) by qualifications

	Physical and Emotional Care		Interactive Care		Travel and Communication		Passive Care	
	Father	Mother	Father	Mother	Father	Mother	Father	Mother
Constant term	12.37	75.61*	34.98***	88.85***	2.23**	12.03***	44.02*	202.02***
Qualifications								
Basic vocational	13.10	20.50	9.07	-9,51	-5.36	-14.98**	35.02	32.57
Skilled vocational	1.44	29.04**	3.09	2.81	1.07	-5.27	12.23	39.19*
University diploma	5.08	28.18**	9.73	2.98	-0.72	3.37	25.29*	65.98***
Bachelor degree	13.13***	13.67**	18.36**	13.04*	2.30	2.45	44.42***	50.53**
Postgraduate	4.67	30.19**	20.55**	26.23*	2.80	-3.32	50.63**	34.22

* P-value<0.05 ** P-value<0.01 *** P-value<0.001

Source: Author's calculations of ABS TUS 1997. The figures in this table are drawn from Table A8.

care, but not housework, than other fathers. For this reason, and because women's child care time also goes up with educational attainment, the results do not mean that higher education levels are associated with greater domestic gender equity overall. Mothers with higher educational levels manage extra amounts of paid work and child care by cutting back unilaterally on their own time in housework.

The gendered time allocation holds broadly true for child care tasks allocation also (see Table 6.4). The greater physical care found at household level to be associated with higher parental education is nearly all being done by mothers. The model predicts an increase in physical care time (of nearly quarter of an hour a day) for fathers with bachelor degrees, but not for fathers with any other level of education.

This suggests that educated men take on a greater share of the hands-on work of child care than men with less education. However, any impact on the gender division of child care time between partners is diminished by the fact that time inputs to physical care are higher for all qualified than unqualified mothers, with the extra time inputs about 20 minutes a day. So increased female inputs associated with education outstrip male, meaning that the children of highly educated parents receive more total care, not that the extra male inputs lead to more gender equity.

Having a bachelor or postgraduate degree is associated with both fathers and mothers doing more interactive care than other parents. This indicates that the higher human capital-fostering child care input found at household level is contributed by both sexes, and suggests that parents with high education levels put more time than other parents into the types of child care that foster the human capital acquisition of their children.

Echoing the household level findings, time in child-related transport and communication does not vary with parental educational attainment, except that the model predicts that women with basic vocational qualifications will spend about 15 minutes a day less in these activities than other women.

Parental passive care time increases in association with higher levels of education. For men, it is passive care time that increases in association with higher education more than with any other child care sub-type. Men with university diplomas, bachelor degrees and postgraduate degrees all spend from nearly half an hour to an hour more than less educated men supervising their children without active involvement. Most educated mothers are not only more active carers of their children, but also find more time to be in the company of their children, available to be called upon, than do other mothers. This applies to mothers at all qualification levels beyond secondary school except those with postgraduate degrees. This most highly educated group of women allocate more time to interactive and physical care than women without tertiary qualifications, and commit most daily time of all female groups to the paid work force, but do not allocate more (or less) time to supervisory child care than women without higher education. These findings fit with those in Chapter 4, which showed that men's child care time is disproportionately allocated to interactive care and passive supervision. The results in this chapter imply that although educated men do spend more time with children that other men, they are not taking over hands-on care from their wives. Educated fathers are not extending the male repertoire to include more types of child care, but doing more of the child care that men normally do. So the findings suggest that care time will be higher in educated households, but do not strongly suggest that this care will be more equitably divided between marital partners.

The amount of time a parent is with children or performing child care out of the presence of their spouse indicates the degree to which they take responsibility for the whole job of child care, rather than merely assisting with tasks delegated to them (see discussion in Chapter 4). This study investigated whether educational attainment is associated with men taking more independent responsibility for child care, or with more gender equity in responsibility taking. All mothers spend more time in sole charge of children than fathers do (see Column 2 of Table 6.5). Women average nearly half of their time with children without their spouse present. No variation in educational level is associated with a change in this percentage except for women with postgraduate qualifications, who spend slightly less (36 per cent) of their time in sole charge of children than unqualified women. This gives a slight indication that highly educated women share time with children with their partners more equitably

Table 6.5   Percentage of mean time spent by fathers and mothers with children doing child care or in sole charge of children by qualifications

	Percentage of time with children without spouse present	
	Father	*Mother*
No post-school qualifications	0.19	0.48
Qualifications		
Basic vocational	0.12	0.48
Skilled vocational	0.15	0.43
University diploma	0.10	0.43
Bachelor degree	0.13	0.47
Postgraduate	0.20	0.36

Source: Author's calculations of ABS TUS 1997.

than other women. However, this is not reflected in the male figures. Between 10 and 20 per cent of male time with children is spent without a spouse present.

Discussion and Conclusion

The child care practices in households in which the parents are highly educated do seem to reflect acceptance of the current normative ideal of intensive parenting. Parents' time devoted to care of their own children is positively associated with educational attainment. This is particularly true of providing more time in hands-on physical care, more time in children's company, and more time in the interactive activities that are most likely to foster human capital development.

However, there is only small indication of increased gender equity in the provision of that care. Men with postgraduate degrees do slightly more physical care, and university educated men do perform more interactive care tasks with their children than other men, but broadly speaking it is mothers who supply the higher allocation to child care found in educated households. Also, though educated men spend longer in child care than less educated men, not much more of it is in sole charge, because the educated women are also spending longer doing child care. At the same time, as human capital theorists would predict, educated women are averaging longer each day in paid work than other women. They also do less housework. This implies that bargaining models, which suggest that women with more personal resources will negotiate to do less in the home, are not helpful in predicting intra-household time allocation to care. The idea that in those situations in which the partners have more equal command of resources the outcomes will be more equal seems more applicable to housework than to caring for children.

Implicit in bargaining theories is the idea that the more power one has in a relationship, the more one will be able to avoid unpleasant tasks (Brines, 1994). The results above suggest that explanatory power of domestic bargaining theories is limited because women do not necessarily avoid care, even when they do command greater personal resources. Acquiring an education and taking up paid work can co-exist with a wish to care more, rather than less. Those women who have a theoretically stronger negotiating position by virtue of their earning capacity do not seem to use it to avoid care. So the gender-neutral implication of bargaining theories, that improving women's personal resources is a route to greater domestic equity, founders. For both sexes, time with children seems particularly highly valued by those who have good market opportunities. This implies that negotiations over cleaning the bathroom are very different from negotiations over child care. For mothers, having an education is not a solution to the incompatible pull to work and push to care – it actually seems to be associated with an accentuation of the difficulty experienced by all women who attempt to fulfil both these roles.

In the next chapter, I look at whether the impact of children on parental time varies with policy and social environment.

Chapter 7

Cross-national Comparison of the Impact of Children on Adult Time[1]

This book has detailed the impact of children on adult time in Australia. In this chapter, I place Australia within an international context to see if differences in the impact of children are associated with variation in policy and social environment.

Background

Countries vary in their attitudes to children and childrearing, and also in the degree to which the responsibility falls to women or is shared by men or by the state. All countries both expect parents to provide care and accept some social responsibility for children (Folbre, 2002). However, there are differences in the extent, both practically and rhetorically, to which children are seen as a private responsibility and pleasure or as a contribution to common prosperity. At one extreme, children are a private good, and any gap in time or money expenditures between parents and the childless, or between mothers and fathers, is not a matter of public concern. Consequences for labour supply and earning capacity are, similarly, a personal issue for parents. An alternative view is that children are most accurately not a private, but a public good. Rather than being consumption items for parents, children are in fact a *product* of family inputs comprising money, time and labour (England and Folbre, 1997). They have an enormous economic worth, but it does not go to the parents who rear them, but to the government, employers, and the whole community (Folbre, 1994a; Chesnais, 1996; Klevmarken, 1999; Crittenden, 2001). The degree to which a country accepts the public good view of children will be reflected in social policy and rates of child-related government spending. Of interest in this chapter is whether these social differences are reflected in how becoming a parent is experienced in daily life.

Whether children are seen as a private responsibility or joint social responsibility will reflect broader attitudes and practices relating to welfare and community, which vary cross nationally. Comparative international research can be regarded as a 'natural experiment' on the effects of policy variation that is not possible from looking at one country at one point in time (Castles, 2002). There is a wide range of social and economic policies that impact upon the family, and a large body of cross-

---

1 Parts of the argument and analysis in this chapter appeared in Craig, Lyn (2006) 'Do Time Use Patterns Influence Fertility Decisions? A cross-national inquiry' *Electronic International Journal for Time Use Research* 3 (1): 60–88.

national research compares public policies of relevance to the family such as the availability of child care, maternity leave, or flexible work arrangements, and public indicators of their effect, such as maternal work force participation (see for example Bradshaw et al., 1993; Gornick et al., 1996; Orloff, 1997; Plantenga and Hansen, 1999; Gornick and Meyers, 2003). How life is lived within the home in terms of the burden of unpaid work and the division of labour has been the subject of much less cross-national investigation. The relatively recent addition of time-use data to the more established social and economic statistics arsenal offers a methodology by which to address this issue directly, and a small body of cross-national research into intra-household time-use has begun to emerge (Bittman, 1999b; Gershuny and Sullivan, 2003; Bonke and Koch-Weser, 2004; Gornick and Meyers, 2004; Pacholok and Gauthier, 2004).

In this chapter I add to this emerging literature by quantifying the effect of children on adult time-use in four countries with contrasting different policy regimes. Using a framework of welfare state typology pioneered by Gosta Esping-Andersen and developed by Walter Korpi, and data from the MTUS World 5 Series, I compare the time impact of children upon adults in Australia with those in three other countries with different approaches to economic, social and family organisation. This is to see whether the time impacts of children and the way they are allocated by gender is affected by policy environment.

I begin with a discussion of welfare regime typology, looking at whether established welfare regime categories can reflect differences in policy relating to how the time effects of raising children are socially assigned. I give a brief description of the policy in each of the categories and discuss the countries I have chosen to represent each one. Next, I set out the way I measure the time impact of children in this cross-national analysis. There are three aspects. The first is the gap between couple parents and non-parents in total paid and unpaid work undertaken, which gives a measure of the total workload associated with becoming a parent. The second is the way that individual parents allocate their time between paid and unpaid work, which shows how possible it is to combine work and family (for both men and women) in each of the four countries. The third is how paid and unpaid work is divided between mothers and fathers, which gives an indication of gender equity in both the total work and the kind of work contributed to the family.

Welfare Regimes, Gender and Parental Care

Esping-Andersen (1990) developed a welfare regime typology that built on both Titmuss' trilogy of residual, industrial-achievement and institutional-redistributive models and Marshall's idea of social citizenship (Marshall, 1950; Titmuss, 1974). Arguing that the way in which social risks are distributed is the defining feature of a welfare state, Esping-Andersen went beyond expenditure as the sole criterion of welfare effort and devised a typology according to how countries drew on the three pillars of social wellbeing – states, markets, and families. He divided western states into a three-way grouping: 'liberal', 'corporatist' and 'social democratic', according to the extent to which social rights permitted people a reasonable living standard independent of market forces (de-commodification).

He described liberal welfare states, exemplified by the USA, Canada, United Kingdom, Australia and New Zealand as market based. In their purest form, liberal welfare states expect their citizens to provide for themselves through work force participation. Welfare support is largely targeted towards those with demonstrated need, through modest means-tested benefits for those who cannot support themselves. The state intervenes only if the market fails. In corporatist states, a category into which he put Germany, Austria, Italy and France, social insurance systems group people in similar occupations. Rights are accrued through contribution, and are attached to class and status, and so differential welfare is a product of group membership. It is not need that dictates whether welfare requirements are met, but whether or not one is a member of the group. In the social democratic model, exemplified by Norway, Finland, Denmark and Sweden, rights accrue from citizenship and there is equality of the highest standard, not of minimum needs. Under this model, people are bound together across social class because all citizens are potential beneficiaries (Esping-Andersen, 1990).

Feminists responded to Esping-Andersen's three-way typology by pointing out that the social risk structure for women is not covered by the concept of de-commodification. The concept assumed a level of work force participation not necessarily achieved by women. It failed to adequately acknowledge the family's place in the provision of welfare and care and how this impacted upon women in different welfare regimes. In addition to the criteria of the necessity to work and not to work, which reflected the situation of males, an essential dimension of social risk for women is the freedom to provide or to not provide caring services. According to many feminist authors, the crucial relationship is not just between paid work and welfare but between paid work, unpaid work and welfare (Orloff, 1993; Siaroff, 1994; Sainsbury, 1996; O'Connor et al., 1999; Korpi, 2000; Arts and Gelissen, 2002; Lewis, 2002; Lewis and Giullari, 2005; Skevik, 2005).

Feminist research into comparative social policy found that the closer the gender dimension is to the analysis, the more the inadequacy of traditional regime theory is exposed (Sainsbury, 1999). Feminists suggested various criteria that would better reflect the position of women within different welfare state regimes. Lewis (1992) argued that it is important to look at the extent to which policy makers in different regimes presume that women would be dependent on male breadwinners. Sainsbury (1996) argued for attention to caring regimes, that is, policies that constitute and structure women's unpaid work. Daly and Lewis (2000) advocated using social care as a critical dimension for analysing welfare state variation (Daly and Lewis, 2000). Several suggested that a differentiating measure should be the capacity of women to form and maintain an autonomous household (O'Connor, 1993; Orloff, 1993; McLaughlin and Glendinning, 1994; Shaver and Burke, 2003).

The feminist criticism led Esping-Andersen to integrate gender issues more centrally into his typology. In his recent work, Esping-Andersen more emphatically states that the components of welfare regimes are not only labour markets and the state, but also the family and that the sum total of societal welfare derives from how inputs from these are combined. 'Welfare states are an inter-causal triad of state, market, and family' (Esping-Andersen, 1999). He incorporates familialisation, the degree to which citizens' welfare depends on family support, as a criterion for categorisation in welfare regime typology. A familialistic system is one in which public policy reflects the expectation that households are the main provider of their members' welfare. A regime that promotes de-familialisation is one in which the burden on direct family welfare provision can be lessened through market or state provision of care and support (Esping-Andersen, 1999).

Social enquiry was historically able to concentrate on the public policy areas of state and market, sidelining families on the assumption that family inputs were both private and stable. This assumption is no longer justified, as it presumed a society that no longer obtains. The welfare state as a particular historical construction of the early to mid-twentieth century catered to a historically specific population with a historically specific risk structure. The labour force was predominantly male, women were at home, and post-Second World War families were stable with high fertility. These factors are no longer the norm, and there is a disjuncture between existing institutional structures and social patterns. As argued throughout this book, one of the results of this is that parenting has become problematised. In some countries, there has been half a sex revolution, in that women are entering the paid work force, but are receiving little extra assistance with domestic responsibilities. In others, the sex revolution is even less advanced. Esping-Andersen argues that the household's traditional caring capacities are eroding and poverty risks are mounting while families are being asked to absorb new labour market risks. He asserts that policy blindness to the world of families can no longer be justified. 'There is currently a social crisis, its locus is the nexus between the family and the labour market and therefore investigation into household economy is central to a sound understanding of post-industrial society' (Esping-Andersen, 1999, p.54).

A Four-way Typology

It is Esping-Andersen's view that the inclusion of familialisation as a criterion fits into his original tripartite typology, and does not necessitate the addition of separate regime types. However, a major difficulty in categorising regimes from a perspective that acknowledges both gender issues and the way child care is distributed, is the enormous range of social factors that impact upon families, and upon the care-giving burden. The complexity of possible impacts has led some researchers to disagree with Esping-Andersen's conclusion that adding the dimension of familialisation does not necessitate extending his original tripartite model, arguing that no simple categorisation of welfare states covers class and gender neatly (Arts and Gelissen, 2002). Nor is there any that explicitly arise from child-related policies, so it is necessary to view this issue through the overlapping but not entirely commensurate prism of gender.

While there is a range of policies that could assist with the care burden, an important distinction is between policies that promote mothers' access to work and policies that discourage women's work force participation. This relates to an old feminist dilemma, identified by Mary Wollstonecraft (1999 (1792)): the problem of whether women should fight for opportunities to participate in public life on equal terms with men, or to fulfil caring roles at home and be valued for those contributions. In short, the dilemma is whether to base citizenship demands on difference from or equality with men (Pateman, 1988; Lister, 1997; Daly and Lewis, 2000; Hobson et al., 2002). Although many commentators regard the equality-difference dichotomy as unfortunate, divisive or misleading (see for example Fraser, 1994; Lister, 1997; Chambers, 2000), this difficult issue is as yet unresolved. Long-standing arguments about, for instance, whether a wage for mothering would value caring or merely enshrine care giving as women's work, are ongoing (Cass, 1994; Knijin, 1994; Lewis, 1997; Hobson et al., 2002; Skevik, 2005).

In an attempt to build on Esping-Andersen's typology but recognise the gender and caring issue more fully, Korpi (2000) suggests a two-dimensional policy conception in terms of the relative stress on unpaid versus paid work. The first policy choice is whether to leave gender issues to family and markets or whether the state should take an active role, and the second policy choice is whether to promote the dual-earner model or breadwinner model. Using these criteria, he identifies four policy models – dual-earner support, general family support divided into two subgroups, and market orientated. Each has a different approach, sometimes articulated and sometimes not, to how the care of children should be shared between state, market and family (Arts and Gelissen, 2002).

Some question the point of the whole project of regime classification, arguing that it may be less useful to distinguish families of nations in terms of social policy models, than to examine empirical variations along separate dimensions (Boje, 1996; Sainsbury, 1996). The categories can be regarded as ideal types only, no country will unambiguously exemplify what is essentially a heuristic model, and countries in any one grouping are in some ways different (Therborn, 1993; Goodin et al., 1999). Any comparison will be very broad-brush. The picture is further complicated by the fact that welfare states are undergoing change, and the boundaries between regime types are blurring (Andinach, 2002; Arts and Gelissen, 2002; Thevenon, 2003). I acknowledge the force of these objections, but cross-national comparisons, though flawed, can still provide a useful basis for investigating the differential effects of policy. I follow Korpi's four-way categorisation into market-orientated, Scandinavian, continental Western European and southern European. Due to space constraints, I can undertake comparison of only one example of each, and have selected Norway, Germany and Italy to compare with Australia.

I now briefly outline the characteristics of each grouping.

Dual-earner support

Countries that adopt the dual-earner support model implicitly acknowledge the public good aspect of children and do not expect women to be solely responsible for their care. The Scandinavian countries comprise this group. They have policy arrangements that pre-emptively socialise the cost of family-hood and explicitly promote women's independence from family obligations. These include public day care services for young children, and generous paid maternity and paternity leave. Nordic governments have a history of social policies aimed at helping people balance their work and family life, which are comparatively generous by OECD standards (OECD, 2003). Public policies intentionally aim to shift provision of care from the unpaid to the paid sector. There is a heavy social service burden, not only to service family needs but also to allow women to choose work rather than household labour. Such comprehensive coverage requires a large workforce to contribute to the high costs, which reinforces the need for women's employment. Gender equality is an explicit (though not yet fully realised) aim (Korpi, 2000). I have chosen Norway as an example of this type of regime.

In Norway, mothers are entitled to 12 months off work with 80 per cent pay or 10 months with full pay. Mothers must take the first six weeks after birth as maternity leave, but after that it is up to the parents to share the remaining leave as they wish. Also, fathers must take at least four weeks leave or else those weeks will be lost for both parents. The paid leave is financed through taxes, so the cost to both parents and employers is minimised. Childless women and mothers have the same workforce participation rate. Norway has a GDP well above the OECD average, a poverty rate much lower than the other countries studied, and somewhat less pronounced income inequality. The child poverty rate is lower than that of the general population, the only country here studied for which that is the case (OECD, 2005).

General family support

Western Europe   Countries that follow the general family support model place a high rhetorical value on children, and have policy arrangements that both assume their care will be assigned to their mothers and encourage this outcome. They regard the family as the primary source of care and welfare, and public policy supports the 'breadwinner-husband-stay-at-home-mother' family model. Indicators of this type of policy regime include cash child allowances, family tax benefits and public day care services for slightly older children, but not for the very young. Family benefits and taxation measures discourage mothers from working. The principle of subsidiarity means that the state will only interfere if family resources are exhausted (Korpi, 2000). Continental European countries are the usual examples of this regime (Esping-Andersen, 1999; Andinach, 2002; Thevenon, 2003), and I have chosen Germany.

Germany is a state in which policy, informed by the idea that children require maternal care, reinforces the breadwinner husband and female caregiver family model (O'Hara, 1998; Abrahamson, 1999; Trzcinski, 2000). The organisation of social institutions reflects this expectation. There is an undersupply of child care places, and fees are very high. The school day usually ends at 1.00 p.m. In 2000, 70 per cent of all German women were employed (with a very high proportion – 85 per cent – working part-time) but the participation rate for mothers is significantly lower (OECD, 2002; Clearinghouse, 2005). The difference between the participation rate of childless German women and of German mothers of two or more children is over 20 percentage points (OECD, 2002). It is less pronounced when there is only one child. German economic indicators are generally sound, with GDP slightly above the OECD average, and poverty rate and income inequality slightly below.

However, child poverty, at 12.8 per cent, is marginally above the OECD average (OECD, 2005).

Southern Europe Korpi (2000) suggests that the general family support model should be subdivided into two. Southern European countries constitute a meaningful fourth category because of the extreme lack of state intervention, which results in an even greater reliance on family resources than in Western Europe. Other comparative researchers support this sub-grouping, with the fourth category variously called Latin Rim, late female mobilisation or Mediterranean (Leibfried, 1992; Siaroff, 1994; Arts and Gelissen, 2002). The distinguishing feature between the two sub-categories of general family support is that, though both assign the care of children to their mothers, the former provides more state assistance to home carers than the latter. The Southern European example investigated here is Italy.

The national constitution of Italy defines the family as a private domain with which the state should not interfere (Hantrais, 1997). Italy's family policies neither facilitate women's workforce participation, nor generously subsidise home care, which means there is very heavy reliance on family resources. There is low female workforce participation, with little difference between the participation of childless women and mothers. There are few opportunities for part time work. The (predominantly private) nurseries charge high child care fees, there are few child care places for under 3 year olds, and daily hours are limited. So for women working full time, public child care is not a ready option (OECD, 2003). Economic indicators are mixed: GDP is above OECD average, but both general and child poverty rate, and income inequality are higher than the average (OECD, 2005). There is a chronic shortage of affordable housing for young people, and many stay at home until well into adulthood (Avdeyeva, 2006).

Market-orientated

In market-orientated regimes (US, UK, Canada, Australia and New Zealand) children are viewed as a private responsibility, and family and child care issues largely left to private arrangements in the market. Also called liberal welfare states, these regimes are theoretically gender-blind, and though they do not actively facilitate women's independence from care, tend to be open to its occurrence through private market arrangements. Gender concerns matter less than the sanctity of the market (Orloff, 1996; Esping-Andersen, 1999; O'Connor et al., 1999; Korpi, 2000). Although some argue that the US is actually the only true example of a liberal welfare state (Castles and Mitchell, 1993), Australia is usually included in this grouping. It conforms to the categorisation in some ways, and deviates from it in others (Bryson, 1992; Castles and Mitchell, 1993; Cass, 1994; Shaver, 1995; O'Connor et al., 1999). The Australian welfare system is very targeted, and there are almost no universal benefits. Welfare entitlements are heavily predicated on workforce participation (O'Connor et al., 1999). Australia historically used the breadwinner model of welfare entitlement in which benefits were directed through the male head of the household. Women's eligibility was as part of the family unit and home production had an implicit value, as was demonstrated by the family wage, intended as sufficient to maintain a man, his wife and three children, introduced in the 1907 Harvester Judgement (Bryson, 1992; Cass, 1994; Shaver, 1995). Australia is a country in which the work-care model of fathers working full-time, and mothers working part-time is widely adopted (OECD, 2002). Australia has no statutory maternity leave, and arrangements for work and care are a matter for private choice, at least rhetorically. However, measures tend to reinforce traditional gender roles (Forssen and Hakovirta, 2000; Charlesworth et al., 2002). In particular, the relatively generous family tax transfers prioritise assistance to single-earner couple families (McDonald, 2004b). Australia has a relatively low female employment rate, even for childless women. It is one of only four countries (with Spain, Ireland and Italy) in which fewer than half of mothers with two or more children are in the paid work force (OECD, 2002). Australia shows a more marked decline in work force participation for mothers than other OECD countries (Whitehouse, 2001; Campbell et al., 2005; Lee, 2005), and the majority of employed Australian mothers work part-time (Charlesworth et al., 2002; Earle, 2002; OECD, 2002).

There are some features of Australian social policy that have led researchers to argue that it should be categorised separately from the groupings above. Antipodean countries redistribute wealth to a greater extent than other liberal states (Castles and Mitchell, 1993), though gaps between rich and poor have grown in recent years. There is fairly substantial government funding for social services including health and education. Extra-household child care arrangements are highly regulated and the standard is comparatively high (Cass, 1994; Brennan, 1998) but it is expensive, and places are insufficient, particularly for under 3 year olds (Orloff, 1996; Pocock, 2003a; Castles, 2004). Historically, Australia has provided comparatively generous support to sole parents who care for their own children (O'Hara, 1998; O'Connor et al., 1999) although recent welfare-to-work initiatives will change this (Brennan and Cass, 2005; Craig, 2005).

Research Focus and Method

The aim of this chapter is to investigate whether the impact of children upon time allocation varies with policy environment. I calculate the time impact of children in three interrelated ways. First, I calculate the gap between couple parents and non-parents in (a) paid and (b) unpaid work undertaken, to get a measure of any extra workload associated with becoming a parent rather than remaining childless. Second, I calculate the relative time allocation between paid and unpaid work for both men and women in each welfare regime, and how this distribution of time to home and employment is affected by becoming a parent. Third, I calculate the relative contribution to unpaid work by men and women in childless couples, and by mothers and fathers, to establish the extent of gender equity in domestic labour, and how this is affected by parenthood. The data, overall method and broad limitations are described in Chapter 2. Specific limitations are described below.

The most central limitation is that each of the regime types discussed above is represented by a single case. A survey with more countries in each of the analytic cells would provide more grounds for a meaningful comparison. Also, although the range and quality of the MTUS is being constantly improved over time, it has internal limitations. It is necessary to sacrifice detail in order to obtain comparability across surveys. Demographic data are limited. In some countries important regional differences may be missed. Most surveys in the MTUS collect information from only one household member. Due to the time it takes to collate the surveys, the latest surveys of some countries were not included at the time of analysis. The MTUS draws on country time-use surveys of different quality, and which use different collection methods and coding. In some activities, differences of coding arise from differences of definition, which creates further comparability problems. This is especially so for child care, as it is very variously defined (Folbre et al., 2005). Also, child care will be underestimated as most of the surveys do not include secondary activity, an essential and time consuming aspect of care (Craig, 2006a; Craig, 2006b). The surveys were not all done in the same year. To minimise the impact of this, countries whose surveys fell within three years of each other were selected for this study. This in turn led to a further limitation – the data are not very recent. These limitations mean that the results of this study should be interpreted with caution, and can be regarded as preliminary only. A more reliable study of greater depth may entail the comparison of individual countries' pre-harmonised time-use surveys and would certainly require the inclusion of many more countries in the sample. Nevertheless, this study allows a broad cross-national comparison across many variables as a starting point for enquiry into this topic.

I conduct a series of multivariate regression analyses. In all models, the dependent variables are time spent in total work (paid and unpaid), and time spent in unpaid work (including child care) only. Unpaid work is a subset of total work. The variables are explained in Chapter 2. Because the MTUS does not record secondary activities, the dependent variables in this chapter are of primary activity only. (Recall that this is in contrast to many of the measures in the previous chapters, which include non-overlapping secondary activity.) The independent variables of interest in the models are nationality, sex, parenthood and interactions between them. The series of regression models become increasingly specified, using a stepwise (forward) method progressively adding nationality, sex, age of youngest child and then the interacted variables. The first model has nationality and the dummy variable 'female', meaning the default category is an Australian male. The second has the same default category, but has separate dummy variables that combine sex and nationality. The third model introduces new dummy variables for 'youngest child is under 5' and 'youngest child is aged 5–11'. The fourth model interacts the nationality, sex and age of youngest child. The intention is to tease out the separate effects of nationality, gender and being a parent.

The models hold constant age, income, level of education, number of children, day of the week and spouse's employment status. When time in paid work is part of the dependent variable, the model also controls for labour force status, with employed full-time as the default category. The aim is to isolate the effect of the policy environment by holding constant regional differences in the sex, age, income, employment status and educational level of populations. The constant term in all four regression models represents an Australian male aged 35–44, with no tertiary education, no children, in the middle 50 per cent of income and whose spouse works full-time. The model specifications are shown in Table A1. The results reported in this chapter are derived from those set out in Table A17.

Results

I briefly summarise the results of the first three models as they relate to the variables of interest, before proceeding to a more detailed discussion of Model 4.

Model 1

Model 1 (which has independent dummy variables for each nationality and a separate dummy variable 'female') shows that there is slight variation cross-nationally in total work as a primary activity for childless people (see Table A17). Referent category Australian men average 9.16 hours a day. Norwegian and Italian men average about 20 minutes less, and German men about 10 minutes more. This variation is largely comprised of time spent in unpaid work. Cross-nationally, being female is associated with a higher total workload than being male. The model predicts that women will average 40 minutes more total work a day than men, which is double the largest cross-national variation between men. The effect of gender on unpaid labour is even more striking. Childless women average four hours a day in unpaid work compared with a male average of one and a half hours.

The results of Model 1 also show that, cross-nationally, the presence of children in a household brings with it a higher workload, an average increase in total work of 21 minutes a day for each child. The average increase in primary unpaid work with each child is just under half an hour a day. That the unpaid work increase is higher than the total shows that households adjust to the presence of children at least partly by reducing the amount of paid labour they supply to the market.

Model 2

Model 2 (which combines sex and nationality in a series of dummy variables) shows that having one independent variable for sex obscures important cross-national variation (see Table A17). There are significant workload differences between women in each of the countries. The cross-national finding in Model 1 that women do more total work than men is shown by this more detailed analysis to reflect the behaviour of Italian women only. Italian women do by far the most work of all groups in the sample. Italian men do by far the least. This means that within Italy, there is a large difference between male and female total workloads, which is not found in the other countries. The finding in Model 1 that women do more unpaid work than men does hold true for each of the separate countries, but the amount is much the higher for Italian women, and much lower for Norwegian women.

Model 3

Model 3 (which contains dummy variables for 'youngest child is under 5' and 'youngest child is aged 5–11') shows that, cross-nationally, the average time demand of parenthood is higher the younger the child (see Table A17). The presence of a youngest child under 5 years old is associated with over 50 minutes a day extra total paid and unpaid work as a primary activity, and the presence of a youngest child aged between 5 and 11 years with 16 minutes. When the dependent variable in Model 3 is unpaid work, the additional time associated with the presence of an under-5-year-old is an hour and 10 minutes a day. The presence of a 5–11-year-old is associated with an extra 21 minutes a day.

Model 4

Having dummies for age and number of children (as in Models 1–3) averages their impact across gender and nationality. Model 4 interacts nationality, gender and the age of the youngest child, and therefore shows the effect of parenthood in each country for each sex, and for each age group of children. These results are now discussed in detail. All the tables below show or are derived from fitted values taken from Table A17.

Total Productive Activity

Table 7.1 shows the number of hours worked by women and men from each country when they are childless (Column 1), have a youngest child under 5 (Column 2), or a youngest child aged 5–11 (Column 3). These figures provide a basis on which to calculate the proportion of time that is spent in total productive activities (total paid and unpaid work) by parents compared to non-parents in each country. This indicates the relative time allocation associated with being a parent and with not being a parent *within* countries.

Table 7.1   Predicted hours a day spent in total productive activity (paid and unpaid work) by nationality and parental status

	No Children		Youngest Child Under 5		Youngest Child 5-11	
	Women	*Men*	*Women*	*Men*	*Women*	*Men*
Nationality						
Italy	10.15	8.11	11.53	9.29	10.65	8.61
Germany	9.26	9.63	10.80	11.09	-	-
Australia	9.16	9.63	10.89	10.30	10.17	10.13
Norway	9.22	9.17	9.89	11.00	9.72	9.67

Source: Author's calculations of MTUS World 5.5. The figures in this table are fitted values drawn from Table A17.

It is in the social democratic country of Norway that the workloads of mothers are most similar to that of childless women. A Norwegian mother of a child under 5 has a workload of 7 per cent more than that of a childless Norwegian woman. It is in the liberal state of Australia that mothers have time commitments that are most different from those of their childless compatriots (about 20 per cent higher). The next highest impact is in corporatist Germany (16 per cent). The country representing the Mediterranean sub-category of corporatist states, Italy, has the second lowest female total time increase associated with parenthood. In Italy a mother's workload is 113 per cent of a childless woman's. Of course, this is added to the higher female workload for childless Italian women who, for this reason, continue to have the highest total workload of all women (indeed, all people) post-parenthood as well as when childless, despite the relatively lower impact of motherhood itself (see Table 7.1).

The effect of parenthood upon men in Norway and in Australia is the opposite of that upon women. It is most pronounced in Norway and least pronounced in Australia. A Norwegian father has a workload that is 20 per cent higher than that of a childless Norwegian man. In Australia the difference between fathers and non-fathers is the least of the countries studied (7 per cent). So of the countries studied, the impact of children upon adult time is most similar by sex in Norway, and least similar by sex in Australia.

Unpaid Work

The impact of becoming a parent upon unpaid work is most pronounced for Australian women and least pronounced for Italian men (see Table 7.2). Italian fathers do much less unpaid work than their counterparts in the other countries. Italian mothers of under-5-year-olds do an hour and a quarter more unpaid work than their childless compatriots. Even though this is added to a very substantial pre-child average, it is not Italian mothers who have the highest unpaid workload, but Australian mothers. Indeed, motherhood is associated with a higher increase in unpaid workload in all the other countries, including Norway, than it is in Italy. The biggest factor impacting on unpaid workload in Italy compared to the other countries is being female, whereas in Australia it is motherhood (see Table 7.2).

In all four countries investigated, parents do a great deal more unpaid work than childless people do (see Table 7.2). The presence of children brings a greater increase in unpaid work than in total work. This shows that although total workload does rise following parenthood, reallocation of time from paid to unpaid work is the most widespread strategy to accommodate the time demands of children. However, the degree to which this is true is not universally consistent. The effect of parenthood upon unpaid work varies with both sex and nationality. With a youngest child under 5, most mothers do nearly twice the amount of unpaid work childless women do. The exception is Italian women, who experience a comparatively low unpaid time impact attendant upon parenthood. Italian mothers do about 40 per cent more unpaid workload of childless Italian women. While still substantial, this is by far the least change in the four countries in the amount of unpaid work associated with motherhood.

Table 7.2   Predicted hours a day spent in unpaid work by nationality and parental status

	No Children		Youngest Child Under 5		Youngest Child 5-11	
	Women	*Men*	*Women*	*Men*	*Women*	*Men*
Nationality						
Italy	4.77	1.05	6.64	2.05	5.18	1.05
Germany	3.92	2.12	6.72	3.46	-	-
Australia	3.79	2.12	7.07	2.73	5.24	2.12
Norway	3.20	2.12	5.93	3.68	3.84	2.12

Source: Authors calculations of MTUS World 5.51. The figures in this table are fitted values drawn from Table A17.

Surprisingly, for men the difference in unpaid work is most between Italian fathers and childless Italian men (195 per cent), although because this is from a very low base, Italian fathers still have the lowest absolute amount of unpaid work in the sample (see Table 7.2). The biggest difference between mothers and non-mothers in the amount of time devoted to unpaid work within one country is in Australia. Conversely, the least difference between fathers and non-fathers in this measure is in Australia. This suggests that it is in Australia that the impact of becoming a parent differs most profoundly by sex.

With a youngest over 5, fathers in no county spend more time in unpaid work than childless men do. Norwegian mothers with a youngest over 5 years old spend 20 per cent more time than their childless compatriots do in unpaid work; Italian women 8 per cent, and Australian mothers 40 per cent.

Division of Domestic Labour

Also of interest is how parenthood affects the division of domestic labour between men and women in each of the countries. Childless Italian women do over four times the amount of unpaid work Italian men do, and German women do nearly twice the unpaid work German men do. In Italy, the presence of young children does not initially deepen the division of unpaid labour. It is already very unequal (see Table 7.2). It actually gets a little better when there is a child under 5. It is still, however, the deepest division of domestic labour cross-nationally, and with a youngest child of school age, the division is deeper still. In such families, Italian mothers are averaging over 450 per cent more time in unpaid work than are Italian fathers.

In all the other countries the gender division of unpaid labour is deepened by the presence of children, but to different degrees. Norway is the most equitable in childless households, with women doing about 50 per cent more unpaid work that men. There is not very much change associated with parenthood (mothers of children under 5 do about 70 per cent more unpaid work than fathers). This is largely because the contribution of Norwegian fathers is high compared to that of fathers in other countries. German ratios of male to female unpaid labour also do not alter very much following parenthood. The impact of parenthood on equity of unpaid work is the most profound in Australia. It is in Australia that parents most differ from childless people on this measure. Childless Australian women do 80 per cent more unpaid labour than childless Australian men do, while Australian mothers with children under 5 do a hefty two and a half times the unpaid labour that fathers of children that age do. Australia also continues to show a relatively big inequity in unpaid work as children mature.

Work–family Balance

I now look at how parenthood affects the allocation of time between paid and unpaid work responsibilities. Table 7.3 shows the proportion of total work time that is paid (for childless people, for parents with a youngest child under 5 and for parents with a youngest child 5–11). This is an indication of how people balance employment and home commitments before and after becoming parents.

There are differences in the extent to which children impact upon proportional allocation to paid and unpaid work both by sex and nationality. In all four countries, men spend a much higher proportion of their total work time in activities for which they are paid than do women (see Table 7.3). This is the case even for the childless, suggesting that, cross-nationally, gender itself has a strong impact on the way people apportion their labour. However, parenthood has an independent additional effect. Fathers spend a higher proportion of their total work time in unpaid labour than childless men do. However, the effect upon men is very modest compared with that upon women. It is also of shorter duration. In all the countries for which data are available, fathers of under-5-year-olds allocate proportionately more time to unpaid labour than do either childless men or fathers of 5–11 year-olds. Indeed, in all countries studied, the latter group has an even lower ratio of unpaid to paid work than childless men.

Having a child in either age group is associated with women spending a lower proportion of their total work time in remunerative labour and a higher proportion in unpaid work, though the impact is weaker as the children mature. Cross-nationally,

Table 7.3   Percentage of men and women's total work time that is paid by nationality

	No Children		Youngest Child Under 5		Youngest Child 5-11	
	Women	*Men*	*Women*	*Men*	*Women*	*Men*
Nationality						
Italy	44	90	32	80	39	93
Germany	50	79	26	69	-	-
Australia	51	76	22	71	35	79
Norway	61	74	34	62	44	78

Source: Author's calculations of MTUS World 5.51.

mothers spend a much higher proportion of their total work time in unpaid labour than childless women do, but the change after becoming a mother does vary between countries. It is in the liberal state of Australia that having a child under 5 is associated with the biggest impact on women's relative time allocation to paid and unpaid work. Childless Australian women spend over half their total work time in paid labour. When Australian women have a child under 5, the average of their total work time that is paid is only 22 per cent. This proportion rises to 35 per cent when the youngest child is aged 5–11. This is the lowest proportion of paid to unpaid work found in my sample. On this measure, the impact of children is in this study most profound upon women in Australia.

The paid to unpaid work ratio of childless German women is slightly lower than for childless Australian women (50 per cent), but the impact of children upon the proportional allocation by German women to paid and unpaid labour is slightly less pronounced than in Australia. It is in the social democratic country of Norway that mothers have the highest proportion of paid to unpaid work of all the mothers in the sample (34 per cent), though it is substantially lower than that of their childless compatriots (61 per cent). It reduces in association with the presence of a youngest child by 27 percentage points and of a youngest child aged 5–11 by 17 percentage points. These are very similar relative reductions to those found in both corporatist Germany, and in liberal Australia, suggesting that the gap in time allocation between childless women and mothers, though different in absolute amount across welfare regimes, remains similar in relative terms.

Unexpectedly, it is in the highly familialistic corporatist state of Italy that having children least alters the relative allocation of female time to paid and unpaid work. Childless Italian women have the lowest proportional allocation of time to paid work (44 per cent), but this goes down by only 8 percentage points to 32 per cent following motherhood. Of all the countries studied, motherhood has the least effect on the relative allocation of paid and unpaid work in Italy. Again this suggests that the time penalty of being female in Italy outweighs that of motherhood.

Speculations on Fertility

The findings give rise to speculation as to whether the time impacts of children encourage or inhibit men and women from becoming parents at all. Birth rates are falling worldwide. This is of concern to policy makers because, inter alia, a below-replacement birth rate means an ageing population in which there are insufficient workers to maintain an adequate tax base (Betts, 1998; Chesnais, 1998; Esping-Andersen, 1999; Barnes, 2001; Beaujot, 2001; McDonald and Kippen, 2001). Some suggest that the falling birth rate results from the difficulties women face in meeting the conflicting demands of work and family (Chesnais, 1996; McDonald, 1997; Bryson et al., 1999; Esping-Andersen, 1999; Weston and Qu, 2001). Whereas historically it was assumed that having children was negatively associated with female work force participation, now the causality appears to run the other way. Fertility rates and female participation and employment have since the 1980s become positively correlated (Chesnais, 1996; Chesnais, 1998; Brewster and Rindfuss, 2000;

Castles, 2002; Bonke and Koch-Weser, 2004). It is suggested that this is particularly relevant to regimes in which childless women have equal educational and work opportunities with males, but in which, following parenthood, women retain the major responsibility for the care of their children (McDonald, 2000; McDonald, 2001; McDonald, 2006). Faced with a choice between work and children, women are increasingly choosing work.

Another suggestion is that the degree to which couples share the burden of children is also critical to sustaining fertility levels (Ronsen, 1998; Henneck, 2003). The findings in Chapter 6 suggest that bargaining theory is of limited use in understanding how the labour of parenthood is divided between men and women. However, the theory may provide some insight into whether people are likely to become parents in the first place. Widmalm (1998) suggests that the failure to strike a fair bargain over the distribution of labour can mean not that certain household tasks will fall by gender default, but will not be undertaken at all. 'In a non-cooperative equilibrium, the household public good produced as a result of domestic labour is under provided' (Widmalm, 1998). The argument is that if the domestic labour of partners is very unequal, women will withdraw their contribution to the provision of household public goods that are created by unpaid work, even at the cost of not having something they value, such as a clean house. Of interest here is whether non-cooperation and withdrawal from household undertakings extends to decisions about having children. In other words, where the domestic division of labour is particularly inequitable, if the household public good 'children' is withheld, fewer children are born and national birth rates are low.

While acknowledging that time impacts of gender and parenthood are only one of many factors that will impact upon fertility decisions, the results of this chapter can provide a speculative test for the contentions above. In 1995, the countries in this study had birth rates as follows: Australia 1.82; Italy 1.18; Norway 1.87; Germany 1.25 (International Clearinghouse, 2005). It is the countries that have the least marked gender division of labour among childless people (Norway and Australia) that have the highest birth rates. Italy has both the most marked gender division of labour and the lowest birth rate. Italy and Norway represent the extremes of birth rates in this study, as well as the extremes of absolute unpaid work and the division of labour. Germany is an exception, but otherwise, the countries in which the gender division of labour is least pronounced have the highest birth rates and the countries in which the gender division of labour is most pronounced have the lowest. Some contend that that low fertility rates are associated with the degree to which becoming a mother erodes women's equality. At first glance, the findings in this study show little support for this view. The time impact of children is highest in relative terms for Australian women (biggest increase in total workload, biggest increase in unpaid work, biggest downward impact on workforce participation, biggest impact on domestic division of labour), but it has the second highest birth rate in the sample. This would seem to imply that this change in circumstances is not an impediment to fertility. However, a closer look at the results does find some support for the idea that that low birth rates may relate to gender inequity. Increasingly few Australian women are having a second child (Kippen, 2001). Italy has a high number of women remaining childless. Having regard to their high total workload, relatively low work force participation and the high domestic time demand upon them, this may be because they are already fully stretched. Also possibly, Australian women are having one child, but limiting subsequent births as a result of the profound impact upon the gender division of labour concomitant with the first. In Australia parenthood is associated with a greater bifurcation of paid and unpaid work than it is in the other countries. This lends tentative, speculative support for the idea that Widmalm's bargaining theory can be related to fertility decisions – that in a non-cooperative equilibrium the household public good (in this case, children) will be under-provided (Widmalm, 1998). It raises the possibility that if women receive insufficient domestic assistance they will limit their fertility. While the present sample is too small to test this proposition, the issue of whether domestic inequity is implicated in fertility decline is worth further investigation. More research, using a larger number of countries, is indicated.

Discussion and Conclusion

In this chapter I investigated the time impact of becoming a parent in four different countries, categorised according to a welfare regime typology in which social and policy approaches to the care of children are at different points along the continuum from private good to public responsibility. In all the countries studied, gender is a far more significant predictor of the amount of work done and the proportion of it that is unpaid than is any variation between nations. In all four countries the presence of children brings extra work, both in total and in the proportion that is unpaid, and the effect is stronger the younger the child. However, there are significant cross-national differences in the degree to which this is so, and in how the unpaid labour associated with children labour is allocated by sex. These differences may be related to the policy regime that pertains.

Typically, market-orientated (liberal) regimes do not actively facilitate women's independence from caring duties or their access to paid work, though they are theoretically gender blind, and open to private market arrangements which could promote these outcomes. In this study, it is the liberal state of Australia in which parenthood brings with it the most change in time allocation for women, in magnitude, accentuation of the gender division of domestic labour, and in the reallocation of time from paid to unpaid work. Dual-earner regimes do actively promote women's access to the workforce both before and after motherhood, parental leave is generous, and care for children is generally state-provided. In this study Norway showed the least marked time impacts upon men and women who become parents, compared to the childless, and the most equitable division of these impacts by sex (though they were by no means completely so). The general family support regime, exemplified by Germany, fell into the middle impact range. In the highly familialistic state of Italy, the time effects of becoming a parent are overshadowed by the influence of gender. The division of domestic labour, and the gap between men and women's total workloads, are by far the most extreme of all the countries looked at, but are not accentuated by parenthood.

This could mean that the impact of children on adult time, and how it is divided by sex, varies with the point a country has reached in the sex revolution, and that this in turn is affected by the extent to which social policy acknowledges that children are a public good. It is in the market-orientated regime, where women have good market opportunities but children are conceptualised as a private good and relatively little social assistance is given for caring, that the time impact of children upon mothers is most profound. This suggests that the stalled sex revolution is most pertinent in this regime (in my example, Australia), in that becoming a parent is accompanied by the most change in lifestyle for women and the least for men. Leaving parental time allocation to private arrangement results in greater inequity for mothers than for childless women. In Norway, becoming a parent has more equal outcomes by sex, and arguably because gender equality is a social goal in dual-earner support countries, and the care of children is more socialised, motherhood does not represent such a change in lifestyle from being childless. At the other extreme, where the sex revolution is least advanced, gender inequity is entrenched and the care of children almost entirely a private female responsibility (Italy), the lifestyle change attendant on parenthood is also not very marked. Both mothers and childless women in Italy shoulder a disproportionately large domestic burden.

While any conclusions must be tentative, because the study is so small, the MTUS data are so limited and because policy and social environments are so multi-factorial, the findings of this study offer insight into how parenthood affects the paid and unpaid workload (and its distribution by gender) of citizens in different countries. These results imply that liberal states are those in which the lifestyle changes attendant upon parenthood fall most heavily, and most inequitably by sex. However, the results also imply that the burdens of being female in a highly familialistic regime are so heavy that relatively few women attempt to add the responsibilities of parenthood. Further research of more countries with improved data would test and build on these early speculative results.

Chapter 8

Conclusion

Having children has in recent years become increasingly problematic, despite its centrality to the human condition. Caring for and raising children has not been adequately integrated into the social shifts of the nineteenth and twentieth centuries, in particular the fact that women earning money now requires them to commit time away from home. When women are housewives time devoted to children is not independent from their other duties. Now that more women are gaining an education and joining the paid labour force, time devoted to child care has become a major discrete cost of raising children. While this has markedly increased the cost to women of having children, there has been inadequate compensatory adjustment in either the public or the private sphere. The labour market has made little accommodation to women's unequal responsibility for children and, while women have gained some domestic gender equity by unilaterally reducing the time they devote to housework, they have not applied the same strategy to child care. This means the sex revolution has stalled mid-cycle. It has foundered on the issue of who takes care of the children.

The impasse is at least partly because caring for children remains largely invisible to social and economic accounting. For several reasons, the issue is not clearly recognised. First, a lack of acknowledgement of the social value of children has sidelined from mainstream enquiry the question of their cost and upon whom it falls. Second, the problem has been made confusing by the widespread elision, both popularly and in social and economic theory, of the issues of gender and parenthood. Third, social research has concentrated on changes in the market sector at the expense of direct investigation into the non-market sector not least because, until the advent of reliable time-use data, there has been a lack of empirical information on how life is lived in the home. As a result, the bedrock social issue of peopling the society is, though highly problematic, also remarkably obscure.

In this book I used time-use data from the ABS TUS 1997 and the MTUS World 5.5, to ameliorate the lack of empirical research into the effects of parenthood on couples and individuals. I undertook a comprehensive and detailed analysis of the impact of children upon adult time and how this impact differs by gender, to quantify the magnitude and distribution of this hidden cost of parenthood. The main purposes of this concluding chapter are to summarise my findings, to consider their implications and to suggest some directions for further research.

First and most basically, I found that the job of child care requires large time investments that result in significant differences between the workload of people who have children and those who do not. The biggest time demand comes with the birth of the first child and the time cost of children is higher the younger they are. Daily workload varies more between non-parents and parents than between parents of differing family sizes. It is the decision to have any children at all that creates the largest time-commitment division. Adults in households with children give up sleep, personal care, paid work and recreation to find time for the unpaid work associated with children. They do much more work overall than people without children.

This means that there are significant differences between parents and the childless in how life is lived. When parenthood is a standard adult experience, this is mainly a life stage difference, but with more people remaining childless, it arguably creates a social divide similar to the more widely recognised sociological schisms between different classes or racial and ethnic groupings. This has implications for social cohesion. That there is decreasing commonality between parents and non-parents is reflected in the growing signs that people are less willing to contribute, through taxes or social accommodation, to the raising of other people's children (Hakim, 2000). The neoclassical economic conception of parenting as a leisure activity, with the logical extrapolation that children are equivalent to pets for whom owners (parents) should be wholly responsible (Folbre, 2001), seems to be gaining popular currency. Signs of this can be detected variously: in the growth of child-free communities; lobby groups for the child free; the lack of safety for children in public spaces; resentment from fellow workers who do not regard responsibilities for children as a legitimate reason why they should cover for parents' absences, and the view that protecting children from everything from cars to trans-fats and unwanted media content is entirely a parental responsibility.

However, any division between parents and non-parents per se is dwarfed by the differential effect parenthood has upon men and women. The impact of children is allocated very inequitably by sex, which means that women's lives are affected by parenthood to a much greater degree than are men's. Specialisation by sex is the most common response to the additional household time demands of parenthood, so the time impact of having children falls much more heavily on women than on men.

Therefore, parenthood has an independent overriding effect that builds upon that of gender. Most obviously, this has a direct effect on women's financial opportunities. If theoretically equal treatment in the labour market is predicated on an infrastructure of domestic difference, the unequal division of family responsibilities will perpetuate occupational segregation and limit women's opportunities to pursue economic citizenship (Cass, 1995, p.54). An inferior market position is both a cause and a result of women's subordinate position in the family. While on average all Australian women shoulder a greater domestic burden and have a smaller proportion of their total work time that is remunerated than men, becoming a mother markedly increases and cements the difference between the sexes. Australian women who individually follow Firestone's (1970) prescription to eschew motherhood do enjoy more equitable outcomes in the gender division of labour. This implies that avoiding motherhood is one way to also avoid the most extreme consequences of gender inequity.

However, there were important dimensions to this that required more detailed investigation. It is well recognised that household labour is invisible to standard accounting. This study found that a great deal of the gendered time impact of children is even more obscure, because it consists of under-recognised dimensions of time commitment such as double activity and different task allocation. Although childless men and women have a total workload that is constituted by very different proportions of paid to unpaid work, they do a similar amount of work in total. Parenthood ends this broad equality, because a great deal of the work of childrearing is done simultaneously with other tasks, and most of this double activity falls to women. By including secondary activity in calculating the time impact of parenthood, this book reveals the extent to which mother's workloads are greater than the workloads of both childless women and fathers.

Simultaneous unpaid work activity that people undertake following the transition to parenthood adds to workload and is not absorbed into total workload by a compensating reduction in other activities. Analysing primary activities shows a simple process of time reallocation, such as time in child care displacing time in personal care. Counting secondary activities shows that parental workloads encroach upon and are performed simultaneously with nominally non-work activities. Some researchers have disputed whether the popular idea that women are working a 'second shift' holds true if the total workloads (paid and unpaid) of males and females are compared. There is broad equality if specialisation is overlooked. However, the findings above mean that there *is* a second shift that reflects extra time inputs and not just task specialisation, but it exists for mothers. Because it is coterminous rather than sequential, it has been invisible both to standard economic enquiry and to simple quantification of time inputs. This is a major reason why the full extent of the constricting impact of becoming a mother is not recognised.

This investigation was able to shed light on another obscure aspect of how parenthood differs by sex, by comparing how men and women in couple households parent not only in absolute but also in relative terms. Mothers not only spend more time with children than fathers do, but they also spend that time differently. Fathers spend a high proportion of their time with children in play activities. Compared to fathers, mothers are disproportionately responsible for the physical care of children. When a mother looks after her children, she is more likely than a father to be doing something else at the same time. Men are less likely than women to do the routine child care tasks that have to be done at a certain time. Women are far more likely than men to be alone with their children. Fathers sacrifice less of their leisure than mothers do, and they help out, rather than take full responsibility for child care.

These results clearly show that the degree of constraint resulting from parenthood, and the experience of caring for or being with children, is not the same for each sex. But neither the amount of work involved, the freedom of choice in when to perform it, nor the differences in the way men and women conduct child care are strikingly obvious, and hence required analysis beyond a simple tally of time spent. I would argue that this is a major reason why the current motherhood predicament is not well recognised. Many of the gendered time impacts identified in this book are not easy to articulate. The ways in which the lives of mothers and fathers are not symmetrical can therefore be invisible to policy makers, non-parents, fathers, and even mothers themselves.

The subtle differences are particularly unlikely to be apparent to policy makers and/or employers. The findings above show that the neoclassical conceptualisation of child care as leisure is a fairly accurate representation of fathering, but not of mothering. Arguably, the idea that caring for children as akin to a hobby, something that is pleasurable and can be done in one's spare time *is* currently true for men. However, it is not true for women. Indeed, it seems likely that the male pattern of child care is only possible because female involvement is more sustained and encompassing. Men can have more discretion to come and go, can do more of the fun tasks and refrain from doing housework at the same time as looking after children precisely because the women are routinely present, do the more arduous child care, and perform the domestic labour while also keeping an eye on the children. This highlights the potential pitfalls of a gender-neutral conception of children. Gender neutrality can mean that women's particular experience is overlooked and male experience is presumed to apply to all. Despite its conceptual inclusion of women, neoclassical economics actually theorises men's experience, and thereby obscures those aspects of life in which men and women's experience does not overlap (Strassman, 1993). If policy makers and employers operate on a theoretical universalisation of fatherhood, the specific needs of mothers are not only likely to be unmet, they are also likely to be unrecognised.

Also arguably, much of the impact is imperceptible to those contemplating it beforehand, so many women who become mothers do so in the absence of full appreciation of the time consequences. This may explain the feelings of shock and bewilderment that new mothers often report. Also, mothers may be unable to fully explain the nuances of constraint and responsibility they feel, and conversely, fathers may be unable to discern the ways in which the time they spend with their children differs qualitatively from that of their wives. These speculations begin to touch on the complicated knot of emotion and practicality at the core of the stalled revolution; the attachment women feel for their own children, and the difficulty of minimising time devoted to their care.

An indication of this is that women prioritise care of their children at considerable cost to themselves. The findings discussed thus far have explored the magnitude, nature and distribution of the time cost of children. I also found that these effects are remarkably similar across different groups of women, even those who are taking up opportunities in the public sphere or are well fitted to do so.

A challenge for women who wish to be both mothers and to access opportunities for paid work is to ensure a high standard of care is still provided to their children. There has been a great deal of concern, not least by mothers themselves, that using extra-household child care will drastically reduce the quantity and quality of parental care. The results of this study suggest that much of the worry that children in day care are missing out on vital parental attention is misplaced. Non-parental child care does not replace mothers' care on an hour-for-hour-basis, and does not reduce time in interactive care (arguably most beneficial to good developmental outcomes for children) at all. This suggests that psychological research findings about children in child care having good outcomes, is possibly not only because substitute care is not in itself harmful to kids, but also because they do not actually get much less parental interaction time. With regard to child welfare, these results are encouraging.

What is at risk, however, is maternal welfare. The findings of this book show that women use non-parental child care as much to reschedule their own care as to replace it. This means that institutional substitution is not a complete answer to how women can manage the demands of work and children. It further implies that working mothers are more willing to contemplate adverse outcomes to themselves than to their children, and are protecting time with children at cost to themselves. They squeeze their own time in non-work activities, and sacrifice leisure and personal care in order to allocate time to children. It appears that these mothers are trying to avoid an unacceptable trade-off between paid work and child care by relinquishing neither and are thereby risking overwork, exhaustion, stress and ill-health.

These findings indicate that the masculinisation of women's work patterns has been concomitant with only very moderate masculinisation of their care responsibilities. I found that this holds true even for women who have better than average market opportunities. While it is well known that educated women are more likely to remain childless, or to have fewer children, I found that those educated women who *do* become parents are actually likely to allocate even more time to child care than other women. They are particularly likely to spend more time in the type of child care that fosters the development of their children's human capital. I also found that, at the same time, educated women average longer each day in paid work than other women. That the best years for establishing a career are also the prime childbearing years creates a time squeeze bottleneck. There is conflict between recouping one's own investment in oneself, and investing in one's children (Joshi, 1998). These findings do not support the idea that domestic inequity results from unequal personal resources, and that women with greater market opportunities will be better able to bargain down their care responsibilities. Indeed, they suggest that educated mothers experience particular difficulty in reconciling the incompatible contemporary imperatives of work and care.

Correspondingly, even in households in which there is actual or potential masculinisation of women's work patterns, feminisation of male care patterns is not very marked. I found that both the use of non-parental child care and high levels of paternal education were associated with slightly more involvement by fathers in the care of their children. But while increased father care is to be welcomed if it brings benefits to children and fathers, as is widely argued, it does not appear to do much to ameliorate the difficulty women have in balancing work and family. Highly educated men do spend more time with children than less educated men, but highly educated women are not much more likely to have men substitute for them than less educated women are. As a society, we are educating our young women in large numbers. The greater choice and opportunity this brings women does not, it appears, include an option for mothers to have their care inputs matched in quantity and kind by their spouse.

Together, these findings imply that female time with children is quite inelastic. Mothers in a variety of situations and with a variety of personal characteristics make time with children a very high priority. It appears that once children are born, it is difficult for mothers to modify the time commitment entailed. To get fairer outcomes and reduce their load, many women have adopted male patterns of housework, but this does not seem to be an option with child care. Even women who have no partner, or use day care, or have better than average market opportunities, maintain similar high time commitments to their children. The results of this study show both that masculinisation of maternal caring patterns has not occurred, and that any project to feminise paternal caring patterns is undeveloped. In summary, this analysis of the time effects of parenthood in Australia found that they are huge, narrowly allocated by gender, obscure and largely inelastic.

However, what the analysis to this point could not show is the degree to which these outcomes are an inevitable, perhaps even desired, part of having children. To test this possibility, I conducted a small cross-national comparison of the magnitude and distribution of the time impacts of parenthood for men and women in Australia, Norway, Germany and Italy. Placing the issue in an international context gave an indication of whether the time effects of having children just come with the territory or are influenced by variations in cultural attitudes and social policy environment. I found that in all four countries the presence of children brings extra work, both in total and in the proportion of that work that is unpaid, and that the effect is stronger the younger the child. The time effects of parenthood seem universal to this extent. However, I also found that there are significant cross-national differences in the extent of the impact, and even more difference in how the labour of parenthood is allocated by sex. This shows that policy and welfare regime context can influence the extent and distribution of the time cost of children.

The small sample limits the strength of the conclusions that can be drawn by relating policy measures to the time impacts I found, but my results do give grounds for speculations that could be explored more thoroughly in later research. They suggest that to minimise the time cost of parenthood, and to make it more equal by sex, requires social measures that go beyond merely allowing women to have market opportunities. It requires institutional support such as affordable and widely accessible child care and generous parental leave, and also supporting men's participation in the home. It was Norway, where such measures are the most extensive, that had the most equitable outcome in the study, both as between parents and the childless, and between mothers and fathers. It was Australia, where childless women's work force participation is high but there is relatively little institutional support for father involvement, no statutory maternity leave and expensive private child care, in which parenthood brought the most pronounced and inequitable changes in female total workload, division of labour and participation in the paid work force. In Italy, where both female workforce participation and institutional support for motherhood is relatively low, there was extreme inequity in the division of labour and in total work done by men and women both before and after parenthood.

So social policy measures can influence the impact of children on adult time and how it is divided by gender. Assumptions that the presence of children has the same implications for women worldwide are clearly wrong. I speculate that the extent to which a regime is willing to institute measures that lessen the load on mothers is informed by the extent to which it is acknowledged that children are a public good. In this small study it is Australian women who are most keenly affected by the stalled revolution in which female market opportunities for childless women are out of step with the domestic demands upon mothers. Pertinently, it is a liberal welfare regime. Where children are primarily regarded as a private good, as in a neoclassical economic framework, the large and highly gendered time impacts of parenthood are less likely to be defined as an appropriate target for social intervention. The theory, which regards people who choose to have children as having accepted the associated costs and benefits, would hold that, by having children, women have revealed a preference for the outcomes described above.

The results in this book show, however, that applying rational choice theories to children and child care has marked limitations. The organising concept of an autonomous individual pursuing personal satisfaction is not applicable to motherhood. The decision to become a parent may be made with inadequate information about the consequences, and the findings here suggest that it is difficult to repudiate or mitigate the commitment once made. After children are born, love, attachment and responsibility kick in, and women will provide care even to their own detriment. It has been suggested that becoming a mother more closely approximates being afflicted with an addiction than exercising a free and ongoing choice (Folbre, 2004). There is a practical temptation for others to exploit this 'addiction'. Acknowledging that children are a joint social concern, that time effects are the major cost of raising them, and that these costs should be shared between all parties who benefit from children rather than disproportionately assigning them to women, would be very costly. If the costs of children are not evenly spread, those who do not contribute have the opportunity to free ride on the effort and investment of those who do. 'Governments would prefer families to pay. Men would prefer women to pay. Non-parents would prefer parents to pay. Employers would prefer workers to pay, and so on' (Folbre, 1994, p.2).

However, while those who do not currently bear the major costs of children might be tempted to leave things as they are, this should be a subject of social policy attention because the present arrangements are unsustainable. In the continued absence of adequate social accommodation of the costs of children, it is likely that more women will choose to limit their families or abstain from having children altogether. The results of this study suggest that women may have reached the limit of their potential to accommodate the demands of parenthood individually. The cost of motherhood is becoming prohibitive. The effects of this go well beyond the personal. 'If social benefits of non-market work devoted to the care of children exceed the personal utility that altruistic providers of this work derive from it, individual choices are not necessarily efficient for society as a whole. As private costs go up, willingness to provide these social externalities goes down' (Folbre, 2002). Many of these externalities (happy children, well-socialised adults) appear to be taken for granted in economic circles. An exception is emerging in the current debate over falling fertility, in which the social benefits of children *are* acknowledged. Among other concerns, a dropping birth rate means there could soon be insufficient workers to maintain an adequate tax base for a rapidly ageing population.

Fertility decisions are complicated, and are influenced by a very wide range of factors (Lesthaeghe, 1998; McDonald, 2001; Weston and Qu, 2001; de Vaus, 2002; Quesnel-Valee and Morgan, 2002; Sleebos, 2003; Cannold, 2005), but the division of domestic labour may have to be added to the list. There may be a relationship between how work is allocated within households and the willingness of women to have children. Of the countries in my sample, those in which the division of domestic labour is most equitable had the highest birth rates. Conversely, where there is extreme inequity even amongst non-parents (as exemplified here by Italy), birth rates are so low as to amount to a baby strike (Hobson et al., 2006). The results of the analysis here need to be tested and extended by further research, but they do imply that if men did more housework, women would have more children.

This suggests a need to extend the repertoire of response to parenthood beyond the existing adjustments by women. As it stands, employers are free riders on women's efforts, while men and the state contribute but not enough. So currently women have three options – not have children, work very hard balancing employment and family, or withdraw from paid work. In other words, as women we can choose between rearing the next generation in our spare time, sacrificing our livelihood to it, or missing out on the experience of motherhood altogether. Some commentary represents these choices as the result of different intrinsic preferences (Hakim, 2000), and of course some women are happy with the situation they are in. But many are simply accepting the best personal fit of the poor alternatives available (Morehead, 2005). Other choices, including having men, employers and the state contribute more to the costs of parenting, cannot be made by women unilaterally. To restart the stalled revolution requires more active participation of the other stakeholders in the parenting project. Women cannot do it alone. 'If the issue of ... shared responsibility for children is avoided, if childrearing becomes only mothers' business, it could be a business with a bleak future' (Joshi, 1998).

References

Abrahamson, P. (1999), *The Male Breadwinner Model under Change: The Case of Germany towards the 21st Century*, Roskilde, Denmark: Roskilde University.

ABS (1998), *Time Use Survey, Australia. Users Guide 1997 Cat No. 4150*, Canberra: Australian Bureau of Statistics.

— (2002), *Child Care Catalogue No. 4402.0*, Canberra: Australian Bureau of Statistics.

— (2005), *Australian Social Trends Catalogue No. 4102.0*, Canberra: Australian Bureau of Statistics.

— (2006), *Barriers and Incentives to Labour Force Participation, Australia, Aug 2004 to Jun 2005* Catalogue No 6239.0, Canberra: Australian Bureau of Statistics.

ACOSS (2000), *Bare Necessities – Poverty & Deprivation in Australia Today*, Canberra: Australian Council of Social Service.

Andinach, M.B (2002), *Women in Work and in the Family: Gender and Reconciliation Practices in Southern Europe*, Florence: European University Institute.

Andorka, R. (1978), *Determinants of Fertility in Advanced Societies*, London: Methuen & Co. Ltd.

Apps, P. and Rees, R. (2000), Household Production, Full Consumption and the Costs of Children, in *Discussion Paper No. 157*, Sydney: Faculty of Law, University of Sydney.

Aries, P. (1962), *Centuries of Childhood*, New York: Vintage Books.

Arts, W. and Gelissen, J. (2002), Three Worlds of Welfare Capitalism or more? A State-Of-The-Art Report, *Journal of European Social Policy*, **12**, 137–158.

Arundell, T. (2000), Conceiving and Investigating Motherhood: the Decade's Scholarship, *Journal of Marriage and the Family*, **62**, 1192–1207. [DOI: 10.1111/j.1741-3737.2000.01192.x]

Astone, N.M., Nathanson, C.A., Schoen, R. and Kim Young, J. (1999), Family Demography, Social Theory, and Investment in Social Capital, *Population and Development Review*, **25**, 1–4. [DOI: 10.1111/j.1728-4457.1999.00001.x]

Avdeyeva, O. (2006), In Support of Mothers' Employment: Limits to Policy Convergence in the EU?, *International Journal of Social Welfare*, **15**, 37–49. [DOI: 10.1111/j.1468-2397.2006.00583.x]

Badinter, E. (1981), *The Myth of Motherhood*, London: Souvenir Press.

Barnes, A. (2001), *Low Fertility: A Discussion Paper*, Canberra: Department of Family and Community Services.

Barrett, M. (1980), *Women's Oppression Today*, London: Verso Editions.

Barten, A.P. (1964), Family Composition, Prices and Expenditure Patterns, in P.E. Hart, G. Mills and J.K. Whitaker, eds, *Econometric Analysis for National Economic Pattern*, London: Butterworth.

Baxter, J. (1993), *Work at Home: The Domestic Division of Labour*, Queensland: University of Queensland Press.

— (2000), The Joys and Justice of Housework, *Sociology*, **34**, 609–631. [DOI: 10.1017/S0038038500000389]

— (2002), Patterns of Change and Stability in the Gender Division of Household Labour in Australia, 1996-1997, *Journal of Sociology*, **38**, 399–424. [DOI: 10.11 77/1440783021287567501]

Baxter, J., Hewitt, B. and Western, M. (2005), Post-familial Families and the Domestic Division of Labour, *Journal of Comparative Family Studies*, **36**, 583–600.

Beaujot, R. (2001). *Earning and Caring: Demographic Change and Policy Implications*. Population Studies Centre, University of Western Ontario Discussion Paper 01-5.

Beck, U. and Beck-Gernsheim, E. (2002), *Individualisation*, London: Sage Publications.

Becker, G. (1965), A Theory on the Allocation of Time, *Economic Journal*, **75**, 493–517. [DOI: 10.2307/2228949]

— (1975), *Human Capital: A Theoretical and Empirical Analysis, with Special Reference to Education*, 2nd edn, Chicago: University of Chicago Press.

— (1981), *A Treatise on the Family*, Cambridge, MA: Harvard University Press.

— (1985), Human Capital, Effort, and the Sexual Division of Labour, *Journal of Labor Economics*, **3**, S33–S58. [DOI: 10.1086/298075]

— (1991), *A Treatise on the Family*, Cambridge, Mass.: Harvard University Press.

— (1994). Human Capital and Poverty Alleviation. *Human Capital Development and Operations Policy Working Papers*.

Beck-Gernsheim, E. (2002), *Reinventing the Family: In Search of New Lifestyles*, Cambridge: Polity Press.

Beggs, J. and Chapman, B. (1988), *The Forgone Earnings from Child-Rearing*, Canberra: Centre for Economic Policy Research, Australian, National University Canberra.

Belsky, J. (2001), Developmental Risks (Still) Associated with Early Child Care, *Journal of Child Psychology and Psychiatry*, **42**, 845–859. [PubMed 11693581] [DOI: 10.1111/1469-7610.00782]

Bergmann, B. (1986), *The Economic Emergence of Women*, New York: Basic Books, Inc.

— (1995), Becker's Theory of the Family: Preposterous Conclusions, *Feminist Economics*, **1**, 141–150.

Berk, S.F. (1985), *The Gender Factory: The Apportionment of Work in American Households*, New York: Plenum.

Betts, K. (1998), Fertility, Migration, and the Ageing of the Population – An Analysis of the Official Projections, *People and Place*, **6**, 33–37. [PubMed 12294854]

Bianchi, S. (2004), Gender and Time: The Subtle Revolution in American Family Life, in *Health and Society Scholars Program*, Philadelphia, PA: University of Pennsylvania.

Bianchi, S., Milkie, M., Sayer, L. and Robinson, J. (2000), Is anyone Doing the Housework? Trends in the Gender Division of Household Labor, *Social Forces*, **79**, 191–228. [DOI: 10.2307/2675569]

Bianchi, S., Robinson, J. and Milkie, M. (2006), *Changing Rhythms Of American Family Life*, New York: Sage.

Bianchi, S.M. (2000), Maternal Employment and Time With Children: Dramatic Change or Surprising Continuity?, *Demography*, **37**, 401–414. [PubMed 11086567]

Bianchi, S.M. and Casper, L.M. (2000), American Families, *Population Bulletin*, **55(4)**, 1–44, Washington, DC: Population Reference Bureau.

Bianchi, S.M. and Robinson, J.P. (1997), What Did you do Today? Children's Use of Time, Family Composition and the Acquisition of Social Capital, *Journal of Marriage and the Family*, **59**, 332–334.

Bittman, M. (1992), *Juggling Time*, Canberra: Australian Governmant Publishing Service.

— (1998), Changing Family Responsibilities: the Role of Social Attitudes, Markets and the State, *Family Matters*, **50**, 31–37.

— (1999a). *Now that the Future has Arrived: A Retrospective of Gershuny's Theory of Social Innovation.* Social Policy Research Centre, Discussion Paper, No.110.

— (1999b), Parenthood Without Penalty: Time Use and Public Policy in Australia and Finland, *Feminist Economics*, **5**, 27–42. [DOI: 10.1080/135457099337798]

— (2004), Parenting and Employment, What Time-Use Surveys Show, pp. 152–170, in Folbre, N. and Bittman, M. eds, *Family Time: The Social Organisation of Care*, London: Routledge.

Bittman, M. and Matheson, G. (1996), All Else Confusion: What Time Use Surveys Show About Changes in Gender Equity, in *SPRC Discussion Paper* No. 72, Sydney: Social Policy Research Centre, UNSW.

Bittman, M. and Pixley, J. (1997), *The Double Life of the Family*, St Leonards: Allen & Unwin.

Bittman, M. and Wajcman, J. (2004), The Rush Hour, The Quality of Leisure Time and Gender Equity, in Folbre, N. and Bittman, M. eds, *Family Time: The Social Organization of Care,* London: Routledge.

Bittman, M., Craig, L. and Folbre, N. (2004), Packaging Care: What Happens When Parents Utilize Non-Parental Child Care, in Folbre, N. and Bittman M. eds, *Family Time: The Social Organization of Care*, London: Routledge, pp. 133–151.

Bittman, M., England, P., Sayer, L., Folbre, N. and Matheson, G. (2003), When Does Gender Trump Money? Bargaining and Time in Household Work, *American Journal of Sociology*, **109(1)**, 186–214. [DOI: 10.1086/378341]

Blau, D.M. (2000), The Production of Quality in Child Care Centres: Another Look, *Applied Developmental Science*, **4(3)**, 136–148.

Blau, F., Ferber, M. and Winkler, A. (1998), *The Economics of Women, Men, and Work*, Upper Saddle River, N.J.: Prentice-Hall.

Blau, F.D., Kahn, L.M. and Waldfogel, J. (2000), Understanding Young Women's Marriage Decisions: The Role of Labour and Marriage Market Conditions, *Industrial and Labor Relations Review*, **53**, 624–647. [DOI: 10.2307/2696140]

Blundell, R. and Lewbel, A. (1991), The Information Content of Equivalence Scales, *Journal of Econometrics*, **50**, 49–68. [DOI: 10.1016/0304-4076%2891%2990089-V]

Boje, T. (1996), Welfare State Models in Comparative Research: Do the Models Describe the Reality?, in Greve, B. ed., *Comparative Welfare Systems: The Scandinavian Model in a Period of Change,* New York, N.Y.: St Martin's Press.

Bojer, H. (2006), Resources Versus Capabilities: a Critical Discussion in Memorandum No. 08/2006. Oslo: Department of Economics, University of Oslo.

Bonke, J. and Koch-Weser, E. (2004), The Welfare State and Time Allocation, Denmark, Italy, France and Sweden, in Giele, J. Z. and Holst, E. eds, *Changing Life Patterns in Western Industrial Societies*, Oxford: Elsevier.

Booth, C., Clarke-Stewart, A., Vandell, D.L., McCartney, K. and Tresch Owen, M. (2002), Child-care Usage and Mother-Infant "Quality Time", *Journal of Marriage and the Family*, **64**, 16–26.

Bourke, J. (1993), *Husbandry to Housewifery: Women, Economic Change and Housework in Ireland, 1890–1914*, Oxford: Clarendon Press.

Bowlby, J. (1953), Some Pathological Processes Set in Train by Early Mother-Child Separation, *Journal of Mental Science*, **99**, 265–272. [PubMed 13053148]

— (1972), *Attachment. Attachment and Loss Volume One*, Middlesex: Penguin Books.

— (1973), *Separation. Attachment and Loss Volume Two*, London: Hogarth Press.

Bradbury, B. (1992), *Measuring the Costs of Children*, Sydney: Social Policy Research Centre, UNSW.

— (1995), *Household Semi-Public Goods and the Estimation of Consumer Equivalence Scales: Some First Steps*, Sydney: Social Policy Research Centre, UNSW.

— (2001), *The Welfare Interpretation of Consumer Equivalence Scales*, Canberra: Centre For Economic Policy Research, Australian National University.

Bradbury, B. and Jantti, M. (1999), *Child Poverty across Industrialised Countries*, Florence: UNICEF International Child Development Centre.

Bradshaw, J., Ditch, J., Holmes, H. and Whiteford, P. (1993), A Comparative Study of Child Support in Fifteen Countries, *Journal of European Social Policy*, **3**, 255–271.

Brennan, D. (1998), *The Politics of Australian Childcare: From Philanthropy to Feminism and Beyond*, Cambridge: Cambridge University Press.

Brennan, D. and Cass, B. (2005), Welfare-to-Work Policies for Sole Parent Families in Australia and the USA: Implications for Parents and Children, in *Australian Social Policy Conference 2005*, Sydney, Australia: University of New South Wales.

Breusch, T. and Gray, E. (2003), A Re-Estimation of Mother's Forgone Earnings Using Negotiating the Life Course (NLC) Data, in *Negotiating the Life Course Discussion Paper Series DP-107*, Canberra: Centre for Social Research School of Social Sciences, The Australian National University.

Brewster, K.L. and Rindfuss, R.R. (2000), Fertility and Women's Employment in Industrialized Nations, *Annual Review of Sociology 2000*, No. 271.

Brines, J. (1994), Economic Dependency, Gender and the Division of Labour at Home, *American Journal of Sociology*, **100**, 652–688.

Brooks-Gunn, J., Duncan, G.J., Klebanov, P.K. and Sealand, N. (1993), Do Neighborhoods Influence Child and Adolescent Development?, *American Journal of Sociology*, **99**, 335–395. [DOI: 10.1086/230268]

Brooks-Gunn, J., Han, W.-J. and Waldfogel, J. (2002), Maternal Employment and Child Cognitive Outcomes in the First Three Years of Life: The NICHD Study of Early Childcare, *Child Development*, **73**, 1052–1072. [PubMed 12146733] [DOI: 10.1111/1467-8624.00457]

Browning, M. (1992), Children and Household Economic Behaviour, *Journal of Economic Literature*, **30**, 1434–1475.

Browning, M. and Lechene, V. (2003), Children and Demand: Direct and Non-Direct Effects, *Review of Economics of the Household*, **1**, 9–31. [DOI: 10.1023/A%3A1021895313920]

Bryant, W.K. and Zick, C.D. (1996a), Are We Investing Less in the Next Generation? Historical Trends in the Time Spent Caring for Children, *Journal of Family and Economic Issues*, **17**, 365–391. [DOI: 10.1007/BF02265026]

— (1996b), An Examination of Parent-Child Shared Time, *Journal of Marriage and the Family*, **58**, 227–237.

Bryson, L. (1992). *Welfare and the State*, Basingstoke and London: MacMillan.

Bryson, L., Strazzari, S. and Brown, W. (1999), Shaping Families: Women, Control and Contraception, *Family Matters*, **53**, 31–38.

Bryson, V. (forthcoming), Time-use Studies: a Potentially Feminist Tool?, *International Journal of Feminist Politics*.

Budig, M.J. and Folbre, N. (2004), Activity, Proximity or Responsibility? Measuring Parental Childcare Time, in Folbre, N. and Bittman, M. eds, *Family Time: The Social Organisation of Care*, London: Routledge, pp. 51–68.

Burgess, A. (1997), *Fatherhood Reclaimed – the Making of the Modern Father*, London: Vermilion (Random House).

Cabrera, N. and Tamis-LeMonda, C. (1999), Perspectives on Father Involvement: Research and Policy, *Social Policy Report*, **13(2)**, 2–27.

Cabrera, N., Tamis-LeMonda, C., Bradley, R., Hofferth, S. and Lamb, M. (2000), Fatherhood in the 21st Century, *Child Development*, **71**, 127–136. [PubMed 10836566] [DOI: 10.1111/1467-8624.00126]

Caldwell, J. (1982), *Theory of Fertility Decline*, London: Academic Press.

— (1999), The Delayed Western Fertility Decline: An Examination of English-Speaking Countries, *Population and Development Review*, **25**, 479–513. [DOI: 10.1111/j.1728-4457.1999.00479.x]

Calhoun, C.A. and Espenshade, T.J. (1988), Childbearing and Wives' Foregone Earnings, *Population Studies*, **42(1)**, 5–37. [DOI: 10.1080/0032472031000143106]

Campbell, I., Chalmers, J. and Charlesworth, S. (2005), The Quality of Part Time Jobs in Australia, paper presented at the *Transitions and Risk: New Directions in Social Policy Conference*, Centre for Public Policy, University of Melbourne, 23–25 February.

Cannold, L. (2005), *What, No Baby? Why Women Are Losing the Freedom to Mother, and How They Can Get It Back*, Fremantle: Curtin University Books.

Caporaso, J.A. and Levinde, D.P. (1992), *Theories of Political Economy*, Cambridge: Cambridge University Press.

Casey, J. (1989), *The History of the Family*, Oxford: Basil Blackwell.

Casper, L. and Bianchi, S. (2002), *Continuity & Change in the American Family*, Thousand Oaks: Sage Publications.

Cass, B. (1994), Citizenship, Work and Welfare: The Dilemma for Australian Women, *Social Politics: International Studies in Gender, State and Society*, **1**, 106−124.

— (1995), Gender in Australia's Restructuring Labour Market and Welfare State, in Edwards, A. and Margery, S. eds, *Women in a Restructuring Australia: Work and Welfare*, Sydney: Allen & Unwin.

Castleman, T. and Reed, R. (2003). '"One day to have my own family…"': Ideas about a successful life from early career professionals in a longitudinal study', in *8th AIFS Conference 12-14 February*. Melbourne.

Castles, F.G. (2002), Three Facts about Fertility: Cross National Lessons for the Current Debate, *Family Matters*, **63**, 23−27.

— (2004). The Future of Families: How Society Chooses Policy and Values, Past and Future, *Eureka Street*, e-journal.

Castles, F.G. and Mitchell, D. (1993), Worlds of Welfare and Families of Nations, in Castles, F.G. ed., *Families of Nations*, Aldershot: Dartmouth Publishing Company.

Chambers, S. (2000), The Cultural Foundations of Public Policy, *Philosophy and Social Criticism*, **26**, 75−81.

Charlesworth, S., Campbell, I. and Probert, B. (2002), *Balancing Work and Family Responsibilities: Policy Implementation Options*, Melbourne: Centre for Applied Social Research, RMIT.

Chesnais, J.-C. (1996), Fertility, Family and Social Policy in Contemporary Europe, *Population and Development Review*, **22**, 729−790. [DOI: 10.2307/2137807]

— (1998), Below-Replacement Fertility in the European Union (EU-15): Facts and Policies, 1960–1997, *Review of Population and Social Policy*, **7**, 83−101.

Clearinghouse International (2005), *The Clearinghouse on International Developments in Youth, Child and Family Policies*, New York: Columbia University.

Cohen, P. and Bianchi, S. (1999), Marriage, Children and Women's Employment: What Do We Know?, *Monthly Labor Review*, **122**, 22−31.

Coleman, M. and Ganong, L., eds (2004), *Handbook of Contemporary Families: Considering the Past, Contemplating the Future*, Thousand Oaks, CA: Sage.

Coltrane, S. (1994), Theorizing Masculinities in Contemporary Social Science, in Brod, H. and Kaufman, M. eds, *Theorizing Masculinities*, Thousand Oaks CA, London, New Delhi: Sage Publications.

— (1996), *Family Man: Fatherhood, Housework and Gender Equity*, New York: Oxford University Press.

— (2000), Research on Household Labor: Modeling and Measuring the Social Embeddedness of Routine Family Work, *Journal of Marriage and the Family*, **62**, 1208−1233. [DOI: 10.1111/j.1741-3737.2000.01208.x]

Coltrane, S. and Adams, M. (2003), The Social Construction of the Divorce 'Problem': Morality, Child Victims, and the Politics of Gender, *Family Relations*, **52(4)**, 363–372. [DOI: 10.1111/j.1741-3729.2003.00363.x]

Cowan, R.S (1976), The Industrial Revolution in the Home: Household Technology and Social Change in the 20th Century, *Technology and Culture*, **17**, 1–23. [DOI: 10.2307/3103251]

— (1983), *More Work for Mother: The Ironies of Household Technology from the Open Hearth to the Microwave*, New York: Basic Books.

Craig, L. (2002), The Time Cost of Parenthood: An Analysis of Daily Workload, in *SPRC Discussion Paper* No. 117, Sydney: Social Policy Research Centre, UNSW.

— (2005), The Money or the Care? A Comparison of Couple and Sole Parent Households' Time Allocation to Work and Children, *Australian Journal of Social Issues*, **40(4)**, 521–540.

— (2006a), Children and the Revolution: A Time-Diary Analysis of the Impact of Motherhood on Daily Workload, *Journal of Sociology*, **42**, 125–143. [DOI: 10.1 177/1440783306064942]

— (2006b), Does Father Care Mean Fathers Share? A Comparison of how Mothers and Fathers in Intact Families Spend Time with Children, *Gender and Society*, **20**, 259–281. [DOI: 10.1177/0891243205285212]

— (2007), Is There Really a 'Second Shift,' and If So, Who Does It? A Time-Diary Investigation, *Feminist Review*, **86**, 1–22.

Craig, L., Bittman, M., Brown, J. and Thompson, D. (2006), *Managing Work and Family*, report prepared for the Department of Family and Community Services and Indigenous Affairs, Australian Commonwealth Government, Canberra.

Crittenden, A. (2001), *The Price of Motherhood*, New York: Henry Holt and Company.

Cross, G. (1993), *Time and Money: The Making of Consumer Culture*, London: Routledge.

Dally, A. (1982), *Inventing Motherhood: The Consequences of an Ideal*, London: Burnett Books Ltd.

Daly, M. (2002), Care as a Good for Social Policy, *Journal of Social Policy*, **31**, 251–270. [DOI: 10.1017/S0047279401006572]

Daly, M. and Lewis, J. (2000), The Concept of Social Care and the Analysis of Contemporary Welfare States, *British Journal of Sociology*, **51**, 281–298. [PubMed 10905001] [DOI: 10.1080/00071310050030181]

Davies, H. and Joshi, H. (1994), Sex, Sharing and the Distribution of Income, *Journal of Social Policy*, **23**, 301–340.

Davies, H., Joshi, H. and Peronaci, R. (2000), Forgone Income and Motherhood: What Do Recent British Data Tell Us?, *Population Studies*, **54**, 293–305.

Davis, K. (1997), Reproductive Institutions and the Pressure for Population, *Population and Development Review*, **23**, 611–625. [DOI: 10.2307/2137577]

de Mause, L. (1974), *The History of Childhood*, New York: Harper & Row.

de Vaus, D. (2002), Fertility Decline in Australia. A Demographic Context, *Family Matters*, **63(Summer)**, 14–21.

Delphy, C. and Leonard, D. (1984), *Close to Home: A Materialist Analysis of Women's Oppression*, Amherst, MA: University of Massachusetts Press.

— (1992), *Gender and Power: Society, the Person and Sexual Politics*, Cambridge: Polity Press.

Dempsey, K. (1997), *Inequalities in Marriage*, Melbourne: Oxford University Press.
— (2001), Women's and Men's Consciousness of Shortcomings in Marital Relations, and of the Need for Change, *Family Matters*, **58**, 58–63.
Deutsch, F. (2000), *Halving it All: How Equally Shared Parenting Works*, Cambridge, MA: Harvard University Press.
Donzelot, J. (1979), *The Policing of Families*, New York: Pantheon Books.
Douthitt, R. (1992), The Inclusion of Time Availability in Canadian Poverty Measures, in ISTAT ed., *Time Use Methodology: Towards Consensus*, Rome, Italy: Istituto Nazionale di Statistica, pp. 83–91.
Duncan, S. and Edwards, R., eds (1997), *Single Mothers in an International Context: Mothers or Workers?*, London: UCL Press.
Earle, J. (2002), Family Friendly Workplaces: A Tale of Two Sectors, *Family Matters*, **61(Autumn)**, 12–17.
Easterlin, R.A. (1973), Relative Economic Status and the American Fertility Swing, in Sheldon, E.B. ed., *Family Economic Behaviour: Problems and Prospects*, Philadelphia: Lippincott.
ECLAC, Economic Commission for Latin America and the Caribbean, Brazilian Institute of Geography and Statistics IBGE and National Statistical Institute of Portugal (1999). Equivalence Scales: A Brief Review of Concepts and Methods. in *Third Meeting of the Expert Group on Poverty Statistics (Rio Group)*, Lisbon: ECLAC.
Ehrenreich, B. and English, D. (1989), *For Her Own Good: 150 Years of the Experts' Advice to Women*, New York: Anchor Books, Doubleday.
Eisenstein, E. (1979), *The Printing Press As An Agent of Change*, Cambridge: Cambridge University Press.
Ekert-Jaffe, O., Gardes, F. and Starzec, C. (2000), Estimating the Cost of Children in Poland Using Panel Data, in *26th General Conference of the International Association for Research in Income and Wealth*, Cracow, Poland.
Ellwood, D. and Jencks, C. (2002), The Spread of Single-Parent Families in the United States Since 1960, in *Conference on the Family and Family Policy*, Maxwell School, Syracuse University.
England, P. (1993), The Separative Self: Androcentric Bias in Neoclassical Assumptions, in Ferber, M. and Nelson, J. eds, *Beyond Economic Man: Feminist Theory and Economics*, Chicago: University of Chicago Press.
England, P. and Folbre, N. (1997), Reconceptualizing Human Capital, in *Annual Meeting of the American Sociological Association*, Toronto, Canada: MacArthur Network on the Family and the Economy.
— (2002), Involving Dads: Parental Bargaining and Family Well-Being, in Tamis-LeMonda, C. and Cabrera, N. eds, *Handbook for Father Involvement: Multidisciplinary Perspectives*, Mahwah, NJ: Lawrence Erlbaum Associates.
— (2003), Contracting for Care, in Ferber, M. and Nelson, J. eds, *Feminist Economics Today: Beyond Economic Man*, Chicago and London: The University of Chicago Press.
England, P. and Kilbourne, B.S. (1990), Markets, Marriages and Other Mates, in Friedland, R. and Robertson, A.F. eds, *Beyond the Marketplace: Rethinking Economy and Society*, New York: De Gruyter.

Espenshade, T. (1984), *Investing in Children: New Estimates of Parental Expenditures*, Washington: The Urban Institute Press.

Esping-Andersen, G. (1990), *The Three Worlds of Welfare Capitalism*, Cambridge: Polity Press.

— (1999), *Social Foundations of Postindustrial Economies*, Oxford: Oxford University Press.

Evans, M.D.R. and Kelley, J. (2002), Changes in Public Attitudes to Maternal Employment: Australia, 1984 to 2001, *People and Place*, **10**, 42−56.

Ferber, M.A. and Nelson, J.A. (1993), *Beyond Economic Man*, Chicago: The University of Chicago Press.

Firestone, S. (1970), *The Dialectic of Sex: The Case for Feminist Revolution*, New York: William Morrow.

Folbre, N. (1991), The Unproductive Housewife: Her Evolution in Nineteenth Century Economic Thought, *Signs*, **6**, 463−484.

— (1994a), Children as Public Goods, *The American Economic Review*, **84**, 86−90.

— (1994b), *Who Pays for the Kids? Gender and the Structures of Constraint*, London: Routledge.

— (1997), Gender Coalitions: Extra-Family Influences on Intra-Family Inequality, in Haddad, L., Hoddinott, J. and Alderman, H. eds, *Intra-household Resource Allocation in Developing Countries: Models, Methods, and Policy*, Baltimore: Johns Hopkins University Press, pp. 263−274.

— (2001), *The Invisible Heart: Economics and Family Values*, New York: The New Press.

— (2002), Disincentives to Care: A Critique of US Family Policy, in *Conference on the Family and Family Policy*, Maxwell School, Syracuse University.

— (2004), Valuing Parental Time: New Estimates of Expenditures on Children in the United States in 2000, in *Supporting Children: English-Speaking Countries in an International Context*, New Jersey: Princeton University.

Folbre, N. and Nelson, J.A. (2000), For Love or Money − Or Both?, *Journal of Economic Perspectives*, **14**, 123−140.

Folbre, N., Jayoung, Y., Kade, F. and Sidle Fuligni, A. (2005), By What Measure? Family Time Devoted to Children in the United States, *Demography*, **42**, 373−390. [DOI: 10.1353/dem.2005.0013]

Forssen, K. and Hakovirta, M. (2000), Family Policy, Work Incentives and Employment of Mothers: Findings from the Luxembourg Income Study, in *The Year 2000 International Research Conference on Social Security*, Helsinki, Finland.

Fraser, N. (1994), After the Family Wage: Gender Equity and the Welfare State, *Political Theory*, **22**, 591−618.

Fukuyama, F. (1999), *The Great Disruption: Human Nature and the Reconstitution of Social Order*, London: Profile Books Ltd.

Furstenberg, F. and Cherlin, A. (1991), *Divided Families: What Happens to Children When Parents Part*, Cambridge, MA: Harvard University Press.

Garfinkel, I. and Haveman, R. (1977), *Earnings Capacity, Poverty and Inequality*, New York: Academic Press.

Gergen, K. (1991), *The Saturated Self – Dilemmas of Identity in Contemporary Life*, New York: Basic Books.

Gershuny, J. (1999). The Work/Leisure Balance and the New Political Economy of Time. *Millenium Lectures, 10 Downing Street Magazine.*

— (2000), *Changing Times: Work and Leisure in Post-Industrial Societies*, Oxford: Oxford University Press.

— (2005), *Busyness as the Badge Of Honor for the New Superordinate Working Class: Working Paper of Institute for Social and Economic Research*, Colchester: University of Essex.

Gershuny, J. and Brice, J. (1994), Looking Backwards: Family and Work 1900 to 1992, in *Changing Households: The British Household Panel Survey 1900–1992*, Colchester: Research Centre on Micro-Social Change.

Gershuny, J. and Sullivan, O. (1998), The Sociological Use of Time-Use Diary Analysis, *European Sociological Review*, **14**, 69–85.

— (2003), Time Use, Gender and Public Policy Regimes, *Social Politics*, **10**, 205–228.

Gershuny, J., Bittman, M. and Brice, J. (2005). Exit Voice and Suffering: Do Couples Adapt to Changing Employment Patterns? *Journal of Marriage and the Family*, **67**, 656–665.

Gershuny, J., Godwin, M. and Jones, S. (1994), Domestic Labor Revolution: A Process of Lagged Adaptation?, in Anderson, M., Bechhofer, F. and Gershuny, J. eds, *The Social and Political Economy of the Household*, Oxford: Oxford University Press.

Gerson, K. (2002), Moral Dilemmas, Moral Strategies, and the Transformation of Gender, *Gender and Society*, **16**, 8–28. [DOI: 10.1177/0891243202016001002]

Giddens, A. (1991), *Modernity and Self-Identity: Self and Society in the Late Modern Age*, Cambridge: Polity Press.

Gilding, M. (1991), *The Making and Breaking of the Australian Family*, St Leonards: Allen & Unwin.

Goldscheider, F. and Waite, L. (1991), *New Families, No Families? The Transformation of the American Home*, Berkeley: University of California Press.

Goodin, R.E., Heady, B., Muffels, R. and Dirven, H.-J. (1999), *The Real Worlds of Welfare Capitalism*, Cambridge: Cambridge University Press.

Gorman, W.M. (1976), Tricks with Utility Functions, in Ardis M.J. and Nobay A.R. eds, *Essays in Economic Analysis: Proceedings of the 1975 AUTE Conference*, Cambridge: Cambridge University Press.

Gornick, J. and Meyers, M. (2003), *Families that Work: Policies for Reconciling Parenthood and Employment*, New York: Russell Sage.

— (2004), Welfare Regimes in Relation to Paid Work and Care, in Giele, J.Z. and Holst, E. eds, *Changing Life Patterns in Western Industrial Societies*, Amsterdam: Elsevier Science Publishing, pp. 45–68.

Gornick, J., Meyers, M. and Ross, K. (1996), Public Policies and the Employment of Mothers, Luxembourg Income Study Working Paper 140.

Gray, M. and Chapman, B. (2001), Foregone Earnings from Childrearing: Changes Between 1986 and 1997, *Family Matters*, **58**, 4–9.

Green, W.H. (2003), *Econometric Analysis*, New York: Macmillan Publishing.

Greenstein, T. (2000), Economic Dependence, Gender and the Division of Labour in the Home, *Journal of Marriage and the Family*, **62**, 322–335. [DOI: 10.1111/ j.1741-3737.2000.00322.x]

Griswold, R.L. (1993), *Fatherhood in America: A History*, New York: Basic Books.

Gronau, R. (1988), Consumption Technology and the Intra-Family Distribution of Resources: Adult Equivalence Scales Re-Examined, *Journal of Political Economy*, **96(6)**, 1183–1205. [DOI: 10.1086/261583]

— (1991), The Intra-Family Allocation of Goods: How to Separate the Adult from the Child, *Journal of Labor Economics*, **9(3)**, 207–235. [DOI: 10.1086/298266]

Hakim, C. (2000), *Work-lifestyle Choices in the 21st Century*, Oxford: Oxford University Press.

Han, W.-J., Waldfogel, J. and Brooks-Gunn, J. (2001), The Effects of Early Maternal Employment on Later Cognitive and Behavioural Outcomes, *Journal of Marriage and the Family*, **63**, 336–354. [DOI: 10.1111/j.1741-3737.2001.00336.x]

Hantrais, L. (1997), Exploring Relationships Between Social Policy and Changing Family Forms within the European Union, *European Journal of Population*, **13(4)**, 339–379. [PubMed 12348440] [DOI: 10.1023/A%3A1005941907983]

Harding, A., Lloyd, R. and Greenwell, H. (2001), *Financial Disadvantage in Australia 1990 to 2000 The Persistence of Poverty in a Decade of Growth*, Canberra: The Smith Family.

Harrington, M. (1998), The Care Equation, *The American Prospect*, **9(39)**, 61–67.

Hartmann, H. (1981), The Unhappy Marriage of Marxism and Feminism: Towards a More Progressive Union, in Sargent, L. ed, *Women and Revolution: A Discussion of the Unhappy Marriage of Marxism and Feminism*.

Haveman, R. and Bershadker, A. (1998), *Self-reliance and Poverty, Net Earnings Capacity versus Income for Measuring Poverty*, Public Policy Brief No. 46, Levy Economics Institute, Bard College, Albany.

Haveman, R. and Buron, L. (1993), Escaping Poverty Through Work: The Problem of Low Earnings Capacity in the United States, 1973–88, *Review of Income and Wealth*, **39(2)**, 141–157.

Haveman, R., Bershadker, A. and Schwabish, J. (2002), The Level and Utilisation of Human Capital in the US, in *Workshop on Social Welfare*, Canberra: Research School of Social Sciences, ANU.

Healey, J., ed. (2002), *Poverty*, Thirroul: Spinney Press.

Henneck, R. (2003), *Family Policy in the US, Japan, Germany, Italy and France: Parental Leave, Child Benefits/Family Allowances, Child Care, Marriage/ Cohabitation and Divorce*, Chicago: Council on Contemporary Families.

Hewlett, S.A. and West, C. (1998), *The War Against Parents: What We Can Do for America's Beleaguered Moms and Dads*, New York: Houghton Mifflin Company.

Hewlett, S.A., Rankin, N. and West, C., eds (2002), *Taking Parenting Public: The Case for a New Social Movement*, Lanham, MD: Rowman & Littlefield Publishers, Inc.

Himmelweit, S. (2000), *Alternative Rationalities, or Why Do Economists Become Parents?*, Milton Keynes: The Open University.

Hobson, B., Lewis, J. and Siim, B. (2002), *Contested Concepts in Gender and Social Politics*, Cheltenham: Edward Elgar Publishing Ltd.

Hobson, B., Olah, L. and Morrissens, A. (2006), The Positive Turn or Birthstrikes? Sites of Resistance to Residual Male Breadwinner Societies and to Welfare State Restructuring, in, *RC19 Meetings of the ISA Sept 2–5*, Paris.

Hochschild, A. (1997), *The Time Bind: When Work Becomes Home and Home Becomes Work*, New York: Henry Holt and Company.

Hochschild, A. and Machung, A. (1989), *The Second Shift*, New York: Viking.

Hofferth, S. (2001), Women's Employment and Care of Children in the United States, in *Women's Employment in a Comparative Perspective*, Van der Lippe, T. and Van Dijk, L. eds, New York: Aldine de Gruyter, pp. 151–174.

Hoffman, L.W. and Youngblad, L.M. (1999), *Mothers at Work: Effects on Children's Well-Being*, Cambridge: Cambridge University Press.

Horrigan, M., Herz, D., Joyce, M., Robison, E.d., Stewart, J. and Stinson, L. (1999), A Report on the Feasibility of Conducting a Time-Use Survey, in *1999 IATUR Conference: the State of Time Use Research at the End of the Century*, Colchester: University of Essex.

Hrdy, S.B. (1999), *Mother Nature: A History of Mothers, Infants & Natural Selection*, London: Chatto & Windus.

Ironmonger, D. (1996), Counting Outputs, Capital Input and Caring Labour: Estimating Gross Household Product, *Feminist Economics*, **2**, 37–64. [DOI: 10.1 080/13545709610001707756]

— (2004), Bringing up Betty and Bobby: The Macro Time Dimensions of Investment in the Care and Nurture of Children, in Folbre, N. and Bittman, M. eds, *Family Time: The Social Organisation of Care*, London: Routledge.

Joshi, H. (1990), The Cash Opportunity Costs of Child Rearing: An Approach To Estimation Using British Data, *Population Studies*, **44**, 41–60. [DOI: 10.1080/00 32472031000144376]

— (1998), The Opportunity Costs of Childbearing: More Than Women's Business, *Journal of Population Economics*, **11**, 161–183. [PubMed 12293833] [DOI: 10.1007/s001480050063]

Joshi, H. and Davies, H. (1999a), The Distribution of the Costs of Children, AEA Session on Maternal Labour Force Participation and Lifetime Earnings. New York.

— (1999b), Measuring the Cost of Children: Estimates for Britain, paper presented at *The Annual Meeting of the International Association for Feminist Economics*, January 12, New York.

Juster, E.T. and Stafford, E.P. (1991), The Allocation of Time: Empirical Findings, Behavioural Models, and Problems of Measurement, *Journal of Economic Literature*, **29**, 471–522.

Kalenkoski, C.M., Ribar, D.C. and Stratton, L.S. (2006), *Parental Child Care in Single Parent, Cohabiting, and Married Couple Families: Time Diary Evidence from the United States and the United Kingdom*, Levy Economics Institute Working Paper No. 440. Available at SSRN: http://ssrn.com/abstract=882083.

Killingsworth, M. and Heckman, J. (1986), Female Labour Supply: A Survey, in Ashenfelter, O. and Layard, R. eds, *Handbook of Labour Economics* Amsterdam: North-Holland.

Kippen, R. (2001), Trends in Age- and Parity-Specific Fertility in Australia, in *International Perspectives on Low Fertility: Trends, Theories and Policies*, Tokyo: Japan: International Union for the Scientific Study of Population Working Group on Low Fertility.

Klevmarken, N. Anders (1999), Microeconomic Analysis of Time-Use Data, Did we Reach the Promised Land?, in Merz, J. and Ehling, M. eds, *Time Use – Research, Data and Policy*, Baden-Baden: Verlagsgesellschaft, pp. 423–456.

Klevmarken, N. Anders, and Stafford, F. (1999). Time Diary Measures of Investment in Young Children, in Smith, J.P. and Willis, R.J. eds, *Wealth, Work and Health. Innovations in Measurement in Social Sciences*, Michigan: University of Michigan Press, pp. 34–63.

Knijin, T. (1994), Fish Without Bikes: Revision of the Dutch Welfare State and its Consequences for the (In)dependence of Single Mothers, *Social Politics: International Studies in Gender, State and Society*, **1**, 83–105.

Korpi, W. (2000). Faces of Inequality: Gender, Class and Patterns of Inequalities in Different Types of Welfare State. *Social Politics: International Studies in Gender State and Society*: 127–191.

Lamb, M., ed. (1997), *The Role of the Father in Child Development*, Somerset, N.J.: John Wiley & Sons, Inc.

Lamb, M.E., Pleck, J.H., Charnov, E.L. and Levine, J.A. (1985), Paternal Behaviour in Humans, *American Zoologist*, **25**, 883–894.

— (1987), A Biosocial Perspective on Paternal Behaviour and Involvement, in Lancaster, J.B., Altman, J., Rossi, A. and Sherrod, L.R. eds, *Parenting Across the Life Span: Biosocial Perspective*, New York: Aldine de Bruyter.

Land, H. (1995), Rewarding Care: A Challenge for Welfare States, in Saunders, P. and Shaver, S. eds, *Social Policy and the Challenges of Social Change: Proceedings of the National Social Policy Conference*, Sydney: Social Policy Research Centre, University of New South Wales.

Land, H. and Lewis, J. (1997), *The Emergence of Lone Motherhood as a Problem in Late Twentieth Century Britain*, London: London School of Economics and Political Science.

Leach, P. (1977), *Baby and Child*, Harmondsworth: Penguin Books.

Leach, P., Sylva, K., Stein, A., Barnes, J. and Malmberg, L.-E. (2005), *Statement on Families, Children and Childcare Study*, London: Institute for the Study of Children, Families & Social Issues.

Lee, C. (2005), Policy, Women's Lives, Women's Futures, *Just Policy*, **4**, 1–16.

Leibenstein, H. (1974), Socio-economic Fertility Theories and Their Relevance to Population Policy, *International Labour Review*, **109**, 443–457. [PubMed 12307190]

Leibfried, S. (1992), Towards a European Welfare State? On Integrating Poverty Regimes into the European Community, in Ferge, Z. and Kolberg, J. eds, *Social Policy in a Changing Europe*, Boulder, CO: Westview.

Lesthaeghe, R. (1998), On Theory Development: Applications to the Study of Family Formation, *Population and Development Review*, **24**, 1–14. [DOI: 10.2307/2808120]

Lewis, J. (1997), Gender and Welfare Regimes: Further Thoughts, *Social Politics: International Studies in Gender, State and Society*, **4**, 160–177.

— (2001), The Problem of Fathers: Policy and Behaviour in Britain, in Hobson, B. ed., *Making Men into Fathers* Cambridge: Cambridge University Press.

— (2002), Gender and Welfare State Change, *European Societies*, **4**, 331–357. [DOI: 10.1080/1461669022000022324]

Lewis, J. and Giullari, S. (2005), The Adult Worker Model Family, Gender Equality and Care: The Search for New Policy Principles and the Possibilities and Problems of a Capabilities Approach, *Economy and Society*, **34**, 76–104. [DOI: 10.1080/0 308514042000329342]

Linder, S. (1970), *The Harried Leisure Class*, New York: Columbia University Press.

Lino, M. (2000), *Expenditures on Children by Families*, Washington, DC: United States Department of Agriculture, Centre for Nutrition Policy and Promotion.

— (2003), *Expenditures on Children by Families, 2002*, Washington, DC: United States Department of Agriculture, Centre for Nutrition Policy and Promotion.

Lister, R. (1997), *Citizenship. Feminist Perspectives*, London: Macmillan Publishing Ltd.

Lovering, K. (1984), Cost of Children in Australia, in Working Paper No. 8: Institute of Family Studies.

Lundberg, S. and Pollack, R.A. (1993), Separate Spheres Bargaining and the Marriage Market, *Journal of Political Economy*, **101**, 988–1010. [DOI: 10.1086/261912]

— (1996), Bargaining and Distribution in Marriage, *Journal of Economic Perspectives*, **10**, 139–158.

Manser, M. and Brown, M. (1980), Marriage and Household Decision-Making: A Bargaining Analysis, *International Economic Review*, **21**, 31–34. [DOI: 10.2307/2526238]

Mattingly, M. and Bianchi, S. (2003), Gender Differences in the Quantity and Quality of Free Time: The U.S. Experience, *Social Forces*, **81**, 999–1030.

Maushart, S. (1997), *The Mask of Motherhood: How Mothering Changes Everything and Why We Pretend It Doesn't*, Sydney: Random House/Australia Pty Ltd.

McDonald, P. (1997), Gender Equity, Social Institutions and the Future of Fertility, in *Working Papers in Demography*, Canberra: The Australian National University.

— (2000), Gender Equity in Theories of Fertility Transition, *Population and Development Review*, **26**, 427–439.

— (2001), Theory Pertaining to Low Fertility, in *International Perspectives on Low Fertility: Trends, Theories and Policies*, Tokyo: International Union for the Scientific Study of Population Working Group on Low Fertility.

— (2004a), The Great Fertility Divide, *Innovation: The Magazine of Research and Technology*, **5**, 42–53.

— (2004b), Work and Family Policies: A Post-2004 Election Review, in *TASA 2004 Conference*, Victoria: La Trobe University, Beechworth Campus.

— (2006), Fertility and the State: The Efficacy of Policy, in *Social Policy in the City Lecture Series*, Social Policy Research Centre, University of New South Wales.

McDonald, P. and Kippen, R. (2001), Labour Supply Prospects in 16 Developed Countries, *Population and Development Review*, **27**, 1. [DOI: 10.1111/j.1728-4457.2001.00001.x]

McElroy, M.B. (1990), The Empirical Content Of Nash − Bargained Household Behaviour, *Journal of Human Resources*, **21**, 333−349.

McElroy, M.B. and Horney, M.J. (1981), Nash Bargained Household Decisions: Toward a Generalisation of the Theory of Demand, *International Economic Review*, **22**, 333−349. [DOI: 10.2307/2526280]

McHugh, M. (1999). *The Costs of Children: Budget Standards Estimates and the Child Support Scheme*, Sydney: Social Policy Research Centre, University of New South Wales.

McLanahan, S. (2002), Life Without Father: What Happens to the Children?, *Context*, **1**, 35−44. [DOI: 10.1525/ctx.2002.1.1.35]

McLaughlin, E. and Glendinning, C. (1994), Paying For Care in Europe: Is There a Feminist Approach?, in Hantrais, L. and Mangen, S. eds, *Family Policy and the Welfare of Women*, Loughborough: University of Loughborough, pp. 52−69.

McMahon, A. (1999), *Taking Care of Men: Sexual Politics in the Public Mind*, Cambridge: Cambridge University Press.

Mein Smith, P. (1997), *Mothers and King Baby: Infant Survival and Welfare in An Imperial World: Australia, 1880−1950*, London: Macmillan.

Meissner, M., Humphreys, E., Meis, S. and Scheu, W. (1975). No Exit for Wives: Sexual Division of Labour and the Cumulation of Household Demands, *Canadian Review of Society and Anthropology*, **12**, 424–439.

Milkie, M., Mattingly, M., Nomaguchi, K., Bianchi, S. and Robinson, J.P. (2004), The Time Squeeze: Parental Statuses and Feelings About Time With Children, *Journal of Marriage and Family*, **66**, 739−761. [DOI: 10.1111/j.0022-2445.2004.00050.x]

Millar, J. and Rowlingson, K., eds (2001), *Lone Parents, Employment and Social Policy: Cross National Comparisons*, Bristol: Policy Press.

Mincer, J. (1962), Labour Force Participation of Married Women: A Study of Labour Supply, in National Bureau of Economic Research ed., *Aspects of Labour Economics*, Princeton, NJ: Princeton University Press, pp. 63−97.

Mincer, J. and Polachek, S. (1974), Family Investments in Human Capital: Earnings of Women, *Journal of Political Economy*, **82**, 76−108. [DOI: 10.1086/260293]

Mitchell, J. (1971), *Woman's Estate*, Harmondsworth: Penguin.

Molm, L. and Cook, K. (1995), Social Exchange and Exchange Networks, in Cook, K., Fine, G. and House, J. eds, *Sociological Perspectives on Social Psychology*, New York: Allyn & Bacon.

Morehead, A. (2005), Governments, Workplaces and Households, *Families Matter*, **70**, 4−9.

Moynihan, D., Rainwater, L. and Smeeding, T.M. (2002), The Challenge of Family System Changes for Research and Policy, in *Conference on the Family and Family Policy*, Syracuse: Maxwell School University.

Myrdal, A. and Klein, V. (1968), *Women's Two Roles: Home and Work*, London: Routledge and Kegan Paul Ltd.

Nava, M. (1983), From Utopian to Scientific Feminism?, Early Feminist Critiques of the Family, in Segal, L. and Harmondsworth, E. eds, *What Is To Be Done About The Family?*, New York: Penguin Books Ltd.

Nelson, J.A. (1992), Methods of Investigating Household Equivalence Scales: An Empirical Investigation, *Review of Income and Wealth*, **38**, 295–310. [DOI: 10.1111/j.1475-4991.1992.tb00427.x]

— (1996). *Feminism, Objectivity and Economics*. London & New York: Routledge.

NICHD, Early Child Care Research Network (1997), The Effects of Infant Child Care on Infant-Mother Attachment Security: Results of the NICHD Study of Early Child Care, *Child Development*, **68**, 860–879. [DOI: 10.1111/j.1467-8624.1997. tb01967.x]

NIOSH (2004), Characteristics of US Workers: Worker Demographics, in *Worker Health Chartbook*, NIOSH Publication No. 2004-146, Washington: National Institute for Occupational Safety and Health.

Nock, S.L. and Kingston, P.W. (1988), Time with Children: The Impact of Couples' Work-Time Commitments, *Social Forces*, **67**, 59–85. [DOI: 10.2307/2579100]

O'Connor, J. (1993), Gender, Class and Citizenship in the Comparative Analysis of Welfare Regimes: Theoretical and Methodological Issues, *British Journal of Sociology*, **44**, 501–518.

O'Connor, J., Orloff, A. and Shaver, S. (1999), *States, Markets, Families*, Cambridge: Cambridge University Press.

O'Hara, K. (1998). *Comparative Family Policy: Eight Countries' Stories*, Canadian Policy Research Networks Study No. F04, Ottawa: Renouf Publishing.

Oakley, A. (1974), *Housewife*, Aylesbury, Bucks: Penguin Books.

— (1979), *Becoming a Mother*, Oxford: Martin Robinson.

— (1981), *Subject Women*, Oxford: Martin Robinson.

— (1985), *The Sociology of Housework*, Cambridge: Basil Blackwell.

OECD (2002), Women at Work: Who Are They and How Are They Faring?, in *Employment Outlook 2002 – Surveying the Jobs Horizon*, Geneva: Organisation for Economic Co-operation and Development.

— (2005), *Society at a Glance: OECD Social Indicators – 2005 Edition*, Geneva: Organisation for Economic Co-operation and Development.

Orloff, A. (1993), Gender and the Social Rights of Citizenship: A Comparative Analysis of Gender Relations and Welfare States, *American Sociological Review*, **58**, 303–328. [DOI: 10.2307/2095903]

— (1996), Gender and the Welfare State, *Annual Review of Sociology*, **22**, 51–78. [DOI: 10.1146/annurev.soc.22.1.51]

— (1997), Comment on 'Gender and Welfare Regimes: Further Thoughts', *Social Politics: International Studies in Gender, State and Society*, **4**, 188–202.

Pacholok, S. and Gauthier, A. (2004), A Tale of Dual-Earner Families in Four Countries, in *Family Time: the Social Organisation of Care*, Bittman, M. and Folbre, N. eds, London: Routledge.

Pahl, J. (1984), *Divisions of Labour*, Oxford: Blackwell.

Pateman, C. (1988), The Fraternal Social Contract, in Keane, J. ed., *Civil Society and the State*, London: Verso.

Perry-Jenkins, M., Repetti, R. and Crouter, A. (2000), Work and Family in the 1990s, *Journal of Marriage and the Family*, **62**, 981–998. [DOI: 10.1111/j.1741-3737.2000.00981.x]

Petre, D. (1998). *Father Time*. Sydney: Pan MacMillan.

Pfau-Effinger, B. (2000), Changing Welfare States and Labour Markets in the Context Of European Gender Arrangements, in Centre For Comparative Welfare State Studies Working Paper.

Plantenga, J. and Hansen, J. (1999), Assessing Equal Opportunities in the European Union, *International Labour Review*, **138**, 351–379.

Pleck, E.H. and Pleck, J.H. (1997), Fatherhood Ideals in the United States, in Lamb, M.E. ed., *The Role of the Father in Child Development*, New York: John Wiley & Sons.

Pocock, B. (2003a), Work and Care: The Australian Response, in IIRA 13th World Congress: Beyond Traditional Employment. Industrial Relations in the Network Economy, Berlin, Germany.

— (2003b), *The Work/Life Collision*, Sydney: Federation Press.

Pollack, R.A. and Wales, T.J. (1979), Welfare Comparisons and Equivalence Scales, *The American Economic Review*, **69**, 216–221.

Powell, L.M. (1997), The Impact of Child Care Costs on the Labour Supply of Married Mothers: Evidence from Canada, *Canadian Journal of Economics*, **30**, 577–594. [DOI: 10.2307/136234]

Presser, H.B. (1995), Are the Interests of Women Inherently at Odds with the Interests of Children or the Family? A Viewpoint, in Mason, K.O. and Jensen, A.-M. eds, *Gender and Family Change in Industrialised Countries*, New York: Clarendon Press, pp. 297–319.

Quesnel-Valee, A. and Philip Morgan, S. (2002), Do Women and Men Realise their Fertility Decisions?, in Annual Meeting of the American Sociological Association, Chicago, Illinois.

Reid, M. (1934), *Economics of Household Production*, New York: John Wiley & Sons, Inc.

Reiger, K. (1985), *The Disenchantment of the Home: Modernising the Australian Family 1880–1940*, Melbourne: Oxford University Press.

Rich, A. (1977), *Of Woman Born*, London: Virago.

Rifkin, J. (1987), *Time Wars*, New York: Henry Holt.

Riley, D. (1983), 'The Serious Burdens of Love' Some Questions on Childcare, Feminism and Socialism, in Segal, L. and Harmonsworth, E. eds, *What Is To Be Done About The Family?* New York: Penguin Books Ltd.

Robinson, J.P. and Godbey, G. (1997), *Time for Life: The Surprising Ways Americans Use their Time*, University Park, PA: Pennsylvania State University Press.

Ronsen, M. (1998), *Fertility and Public Policies − Evidence from Norway and Finland*, Oslo: Statistics Norway.

Rothbarth, E. (1943), A Note on a Method of Determining Equivalent Income for Families of Different Composition, in Madge, C. ed., *War-time Pattern of Saving and Expenditure*, Cambridge: Cambridge University Press.

Russell, G., Barclay, L., Edgecombe, G., Donovan, J., Habib, G., Callaghan, H. and Pawson, Q. (1999), *Fitting Fathers into Families: Men and the Fatherhood Role in Contemporary Australia*, Canberra: Department of Family and Community Services.

Rybczynski, W. (1991), *Waiting for the Weekend*, New York: Viking.

Sainsbury, D. (1996), *Gender, Equality and Welfare States*, Cambridge: Cambridge University Press.

Samuelson, P. and Nordhaus, W. (1985), *Economics*, New York: McGraw-Hill/Book Company.

Sandberg, J. and Hofferth, S. (2001), Changes in Children's Time with Parents: United States, 1981–1997, *Demography*, **38**, 423−436. [PubMed 11523269]

Saunders, P. (1998a), Household Budget and Income Distribution Over the Longer Term, in *SPRC Discussion Paper No 89*, Sydney: Social Policy Research Centre, University of New South Wales.

— (1998b), *Using Budget Standards to Assess the Well-Being of Families*, Sydney: Social Policy Research Centre, University of New South Wales.

— (1999), Budget Standards and the Cost of Children, *Family Matters*, **53(Winter)**, 62−70.

Sayer, L., Bianchi, S. and Robinson, J. (2004), Are Parents Investing less in Children? Trends in Mothers and Fathers Time with Children, *American Journal of Sociology*, **110**, 1−43. [DOI: 10.1086/386270]

Scanzoni, J. (1979), A Historical Perspective on Husband–Wife Bargaining Power, in Levinger, G. and Moles, O. eds, *Divorce and Separation*, New York: Basic Books, pp. 10−36.

Schor, J.B. (1991), *The Overworked American: The Unexpected Decline of Leisure*, New York: Basic Books.

Schultz, T.W. (1974), Fertility and Economic Values, in Schultz, T.W. ed., *Economics of the Family*, Chicago: University of Chicago Press.

Scott, E., Edin, K., London, A. and Mazelis, J.M. (1999), *My Children Come First: Welfare-Reliant Women's Post-TANF Views of Work-Family Tradeoffs, and Marriage*, Chicago: Northwestern University/University of Chicago Joint, Center for Poverty Research.

Seltzer, J. and Brandreth, Y. (1994), What Fathers Say About Involvement With Children After Separation, *Journal of Family Issues*, **15(1)**, 49–77.

Sen, A. (1998), *The Standard of Living*, Cambridge: Cambridge University Press.

Seneca, J.J. and Taussig, M.K. (1971), Family Equivalence Scales and Personal Tax Exemptions for Children, *Review of Economics and Statistics*, **53**, 253−262. [DOI: 10.2307/1937969]

Shaver, S. (1995), Women, Employment and Social Security, in Edwards, A. and Margery, S. eds, *Women in a Restructuring Australia. Work and Welfare*, Sydney: Allen and Unwin.

Shaver, S. and Burke, S. (2003), Welfare States and Women's Autonomy: A Thought Experiment, in *The Australian Social Policy Conference July 9–11 2003*, Sydney: University of New South Wales.

Shelton, B.A. (1992), *Women, Men and Time: Gender Differences in Paid Work, Housework and Leisure*, New York: Greenwood Press.

Shonkoff, J.P., and Phillips, D.A. eds (2000), *Neurons to Neighborhoods: The Science of Early Childhood Development*, Washington, DC: National Academy Press.

Shorter, E. (1977), *The Making of the Modern Family*, London: Fontana Books.

Siaroff, A. (1994), Work, Welfare and Gender Equality: A New Typology, in Sainsbury, D. ed., *Gendering Welfare States*, London: Sage, pp. 82–100.

Silverstein, L.B. and Auerbach, C.F. (1999), Deconstructing the Essential Father, *American Psychologist*, **54**, 397−407. [DOI: 10.1037/0003-066X.54.6.397]

Skevik, A. (2005), Women's Citizenship in the Time of Activation: The Case of Lone Mothers in 'Needs-Based' Welfare States, Social Politics: International Studies in *Gender State and Society*, **12**: 42−66.

Sleebos, J.E. (2003). Low Fertility Rates in OECD Countries: Facts and Policy Responses. *OECD Social, Employment and Migration Working Papers* 15.

South, S. and Spitze, G. (1994), Housework in Marital and Non-Marital Households, *American Sociological Review*, **59**, 327−347. [DOI: 10.2307/2095937]

Spain, D. and Bianchi, S. (1996), *Balancing Act: Motherhood, Marriage and Employment Among American Women*, New York: Russell Sage Foundation.

Spock, B. (1998), *Dr Spock's Baby and Childcare*, 7th edn, New York: Pocket.

Strassman, D. (1993), Not a Free Market: The Rhetoric of Disciplinary Authority in Economics, in Ferber, M. and Chicago, J.N. eds, *Beyond Economic Man: Feminist Theory and Economics*, Chicago: The University of Chicago Press.

Sullivan, O. (1997), Time Waits For No Wo(man): An Investigation of the Gendered Experience Of Domestic Time, *Sociology*, **31**, 221−239. [DOI: 10.1177/003803 8597031002003]

Therborn, G. (1993), The Politics of Childhood: The Rights of Children in Modern Times, in Castles, F.G. ed., *Families of Nations*, Aldershot: Dartmouth Publishing Company Ltd.

Thevenon, O. (2003), *Welfare State Regimes and Female Labour Supply in a European Perspective: A Comparison of Female Behaviour in Germany, France, the Netherlands, Italy, Spain and UK During the 1990s*, paper presented to the Society for the Advancement of Socio-Economics, Aix-en-Provence.

Thompson, L. (1991), Family Work: Women's Sense of Fairness, *Journal of Family Issues*, **12**, 181−196.

Trzcinski, E. (2000), Family Policy in Germany: A Feminist Dilemma?, *Feminist Economics*, **6**, 21−44. [DOI: 10.1080/135457000337651]

Waldfogel, J. (1997), The Effect of Children on Women's Wages, *American Sociological Review*, **62**, 209−217. [DOI: 10.2307/2657300]

— (1998a), The Family Gap for Young Women in the United States and Britain: Can Maternity Leave Make a Difference?, *Journal of Labor Economics*, **16**, 505−545. [DOI: 10.1086/209897]

— (1998b), Understanding the 'Family Gap' in Pay for Women with Children, *The Journal of Economic Perspectives*, **12**, 137−156.

Waller, R. (2002), The Effects of Gender and Dependency on the Division of Household Labour, *Economic Times and Review*, **1**, 1−5.

Waring, M. (1988), *Counting for Nothing*, Wellington: Bridget Williams Books.

West, C., and Zimmerman, D. (1981), Doing Gender, *Gender and Society* **1**: 125–151.

Weston, R. and Qu, L. (2001), Men's and Women's Reasons For Not Having Children, *Family Matters*, **58**, 10−15.

Whitehouse, G. (2001), Industrial Agreements and Work/Family Provisions: Trends and Prospects Under Enterprise Bargaining, *Labour and Industry*, **12**, 109−130.

Widmalm, F. (1998), Marriage, Housework and Fairness, in *Working Paper Series No 22*, Stockholm: Department of Economics, Uppsala University.

Williams, F. (2006), Intersecting Issues of Gender, 'Race', and Migration in the Changing Care Regimes of United Kingdom of Great Britain and Northern Ireland, Sweden and Spain, in *Symposium on Gender and Social Policy*, Sydney: University of Sydney.

Williams, J. (2001), *Unbending Gender: Why Family and Work Conflict and What To Do About It*, Oxford: Oxford University Press.

Wollstonecraft, M. (1999), (1792), *A Vindication of the Rights of Woman*, New York: Bartleby.com.

Wooldridge, J.M. (2003), *Introductory Econometrics*, 3rd Edition, Sydney: South-Western Thomson Learning.

Woolley, F.R. (1996), Getting the Better of Becker, *Feminist Economics*, **2**, 114–120.

Yeung, J., Hill, M. and Duncan, G.J. (2000), Putting Fathers Back into the Picture: Parental Activities and Children's Adult Outcomes, *Marriage and Family Review*, **29**, 97–113. [DOI: 10.1300/J002v29n02_07]

Yeung, J., Sandberg, J.F., Davis-Kean, P. and Hofferth, S. (2001), Children's Time With Fathers in Intact Families, *Journal of Marriage and the Family*, **63**, 136–154. [DOI: 10.1111/j.1741-3737.2001.00136.x]

Young, M. and Wilmott, P. (1973), *The Symmetrical Family: A Study of Work and Leisure in the London Region*, London: Routledge.

Zelizer, V.A. (1985), *Pricing the Priceless Child*, New York: Basic Books, Inc.

Zick, C.D. and Keith Bryant, W. (1996), A New Look at Parents' Time Spent in Child Care: Primary and Secondary Time Use, *Social Science Research*, **25**, 260–280. [DOI: 10.1006/ssre.1996.0012]

Zick, C.D., Keith Bryant, W. and Osterbacka, E. (2001), Mothers' Employment, Parental Involvement and the Implications for Intermediate Child Outcomes, *Social Science Research*, **30**, 25–49. [DOI: 10.1006/ssre.2000.0685]

Appendix

Table A1 Model specifications

Independent variables	
Number and age of children	No children (omitted category)
	Youngest 0-2
	1 child (omitted category when parents only)
	2 children (yes=1)
	3+ children (yes=1)
	Youngest 3-4
	1 child (yes=1)
	2 children (yes=1)
	3+ children (yes=1)
	Youngest 5-11
	1 child (yes=1)
	2 children (yes=1)
	3+ children (yes=1)
Age of youngest child	
	0–2 years (omitted category)•
	3–4 years (yes = 1)•
	0–4 years (yes=1) •
	5–11 years (yes=1)•
Number of children in household	
	One child (omitted category)•
	Two children (yes = 1)•
	Three or more children (yes = 1)•
	No of children under 18♦
Family type	
	Married/de facto (omitted category)*
	Single parent household (yes = 1)*
	Single woman with no children (omitted category)↑
	Single man with no children (omitted category)↑
	Mother with youngest child under 5 (yes = 1)↑
	Mother with youngest child aged 5 to 11 (yes = 1)↑
	Partnered woman with no children (yes = 1)↑
	Partnered man with no children (yes = 1)↑
	Partnered mother with youngest child under 5 (yes = 1)↑
	Partnered father with youngest child under 5 (yes = 1)↑
	Partnered mother with youngest child aged 5 to 11 (yes = 1)↑
	Partnered father with youngest child aged 5 to 11 (yes = 1)↑
Non-parental childcare	
	No non-parental care (omitted category)
	Both formal and informal (yes= 1)†
	Formal (yes = 1)†
	Informal (yes = 1)†
Hours per week child attends day care	Midpoint of values, yields ranges 0-50

Contemporary Motherhood

Table A1   Continued

Independent variables

Qualifications of parent (s)

    No post-school qualifications (omitted category)
    One partner has vocational qualifications, other partner has
      no qualifications (yes = 1)♣
    Both partners have vocational qualifications (yes = 1)♣
    One partner is university educated, the other partner has
      vocational qualifications, or no qualifications (yes = 1)♣
    Both partners are university educated (yes = 1)♣
    Neither partner has qualifications (yes = 1)♣
    Basic vocational qualifications (yes = 1)
    Skilled vocational qualifications (yes = 1)
    University diploma (yes = 1)
    Bachelor degree (yes = 1)
    Postgraduate qualifications (yes = 1)
    Has university qualification (yes=1)↓
    Has vocational qualification (yes=1)↓
    Above secondary♦
    Secondary only (omitted category)♦

Country

    Australia (omitted category)♦
    Norway (yes = 1)♦
    Germany (yes = 1)♦
    Italy (yes = 1)♦

Sex and country

    Female Australian (yes = 1)♦
    Female German (yes = 1)♦
    Female Norwegian (yes = 1)
    Female Italian (yes = 1)♦
    Male Australian (omitted category)♦
    Male German (yes = 1)♦
    Male Norwegian (yes = 1)♦
    Male Italian (yes = 1)♦

Sex, nationality and age of youngest child

    Female Australian and youngest child under 5 (yes=1)♦
    Female German and youngest child under 5 (yes=1)♦
    Female Norwegian and youngest child under 5 (yes=1)♦
    Female Italian and youngest child under 5 (yes=1)
    Male Australian and youngest child under 5 (omitted
      category)♦
    Male Australian and youngest child under 5 (yes=1)♦
    Male German and youngest child under 5 (yes=1)♦
    Male Norwegian and youngest child under 5 (yes=1)♦
    Male Italian and youngest child under 5 (yes=1)♦
    Female Australian and youngest child aged 5 to 11 (yes=1)♦
    Female German and youngest child aged 5 to 11 (yes=1)♦
    Female Norwegian and youngest child aged 5 to 11 (yes=1)♦
    Female Italian and youngest child aged 5 to 11 (yes=1)♦
    Male Australian and youngest child aged 5 to 11 (yes=1)♦
    Male German and youngest child aged 5 to 11 (yes=1)♦
    Male Norwegian and youngest child aged 5 to 11 (yes=1)♦
    Male Italian and youngest child aged 5 to 11 (yes=1)♦

Table A1   Continued

Independent variables	
Controls	
Labour force status	
	Both partners work full time (yes = 1)♣
	Husband works full time, wife not employed (yes = 1)♣
	Husband works full time, wife works part time (omitted category)♣
	Husband not in full time work, other arrangements (yes = 1)♣
	Full time (yes = 1)(omitted category in female only model)
	Part time (yes = 1)(omitted category in male only model)
	Not in the labour force (yes = 1)
Age of respondent	
	Aged 35-44 (omitted category)
	Aged 25-34 (yes=1)
	Aged 45-54 (yes=1)
	Wife is aged 25-34 (yes=1)♣
	Wife is aged 45-54 (yes=1)♣
	Husband is aged 25-34 (yes=1)♣
	Husband is aged 45-54 (yes=1)♣
Day of the week	
	Weekday (omitted category)
	Both diary days weekend (yes =1)♣
	Diary from Saturday and weekday (yes =1)♣
	Diary from Sunday and weekday (yes =1)♣
	Diary day is Saturday (yes =1)
	Diary day is Sunday (yes =1)
Household income	
	Midpoint of ranges, yields values 0-2,300
	Lowest quartile♦
	Middle 50 percent (omitted category)♦
	Highest quartile♦
Disabled person in household	
	No disabled person in household (omitted category)
	Disabled person in household (yes=1)
Hours per week in paid work♠	Midpoint of ranges, yields values 0 – 50
Spouses' hours per week in paid work	Midpoint of ranges, yields values 0 – 50

KEY:
♣ Household level analysis only
♠ Not in model when paid work is the dependent variable
• Only in model used in Chapters 5-6
† Only in model comparing use of non-parental day care type
↓Only in model used in Chapter 5
↑Only for comparing sole parents with whole sample
*Only for comparing sole mothers with partnered mothers
♦Only in model used in Chapter 7

Reference category for households is childless couples aged 35-44, with no post-school education, men work full time and women work part time, use no extra-household childcare, have no disabled household member, on a weekday.

Reference category for individuals is aged 35-44, with no children, no post-school education, men work full time and women work part time, use no extra-household childcare, have no disabled household member, on a weekday.

Table A2   Coefficients of couples' joint hours a day in child care, unpaid work and total work as a primary activity

	Childcare	Unpaid Work	Total Work
Constant term	-0.149***	5.941***	15.538***
	(0.214)	(0.417)	(0.502)
Number and age of children			
Youngest 0-2			
1 child	4.710***	5.056***	2.085***
	(0.227)	(0.443)	(0.586)
2 children	5.612***	6.546***	4.157***
	(0.229)	(0.447)	(0.578)
3+ children	5.195***	6.161***	3.029***
	(0.234)	(0.457)	(0.599)
Youngest 3-4			
1 child	3.181***	4.802***	3.305**
	(0.409)	(0.797)	(1.076)
2 children	4.328***	4.794***	3.099***
	(0.281)	(0.548)	(0.725)
3+ children	3.291***	4.060***	2.385**
	(0.304)	(0.593)	(0.788)
Youngest 5-11			
1 child	1.418***	2.251***	1.244*
	(0.212)	(0.413)	(0.557)
2 children	2.044***	2.056***	2.270***
	(0.166)	(0.324)	(0.433)
3+ children	2.566***	3.362***	2.379***
	(0.229)	(0.447)	(0.602)
Age of parents			
Wife is aged 25-34	-0.004	-0.787**	0.324
	(0.146)	(0.284)	(0.387)
Wife is aged 45-54	0.036	0.280	0.361
	(0.157)	(0.307)	(0.417)
Husband is aged 25-34	-0.133	-0.493	0.067
	(0.150)	(0.292)	(0.395)
Husband is aged 45-54	-0.149	0.124	-0.306
	(0.147)	(0.287)	(0.390)
Qualifications of parents			
Both partners are university educated	0.487***	0.415	1.249**
	(0.149)	(0.290)	(0.394)
Both partners have vocational qualifications	0.278	0.561	1.268**
	(0.166)	(0.324)	(0.440)
One partner has vocational qualifications, other has no qualifications	0.208	0.459	-0.993
	(0.268)	(0.524)	(0.710)
One partner is university educated, other has vocational qualifications,	0.384	0.349	0.313
	(0.198)	(0.387)	(0.526)

Table A2   Continued

	Childcare	Unpaid Work	Total Work
One partner is university educated, other has no qualifications	0.076	0.143	0.833
	(0.154)	(0.301)	(0.409)
Labour Force Status			
Both partners work full time	-0.185	-1.202***	
	(0.143)	(0.278)	
Husband works full time, wife not employed	0.310	0.772**	
	(0.147)	(0.288)	
Husband not in full time work, other arrangements	0.253	0.609	
	(0.160)	(0.312)	
Household income	0.000	0.000	0.002***
	(0.000)	(0.000)	(0.000)
Disabled person in household	-0.012	0.342	-0.400
	(0.113)	(0.221)	(0.299)
Hours per week child attends day care	-0.027***	-0.047***	0.001
	(0.006)	(0.012)	(0.016)
Day of the week			
Both diary days weekend	-0.032	2.143***	-6.530***
	(0.154)	(0.300)	(0.407)
Diary from Saturday and weekday	-0.017	1.207***	-3.917***
	(0.152)	(0.296)	(0.403)
Diary from Sunday and weekday	-0.110	1.043***	-3.720***
	(0.148)	(0.289)	(0.392)
R Square	.600	.404	.320

* P-value<0.05 ** P-value<0.01 *** P-value<0.001   N=1210 . Std error in brackets
Source: ABS TUS 1997.

Table A3   Coefficients of couples' joint hours a day in paid work, personal care and recreation and leisure as a primary activity

	Paid Work	Personal Care	Recreation/ Leisure
Constant	9.090***	22.236***	7.458***
	(0.557)	(0.319)	(0.418)
Number and age of children			
Youngest 0-2			
1 child	-3.740***	-0.870***	-1.263**
	(0.650)	(0.339)	(0.444)
2 children	-3.031***	-1.767***	-2.547***
	(0.642)	(0.342)	(0.448)
3+ children	-3.969***	-2.143***	-1.393**
	(0.665)	(0.350)	(0.458)
Youngest 3-4			
1 child	-1.988	-1.692**	-1.374
	(1.194)	(0.610)	(0.799)
2 children	-2.734**	-1.526***	-2.302***
	(0.805)	(0.419)	(0.549)
3+ children	-2.599**	-1.967***	-1.684**
	(0.874)	(0.454)	(0.594)
Youngest 5-11			
1 child	-1.450**	-0.785**	-1.015**
	(0.619)	(0.316)	(0.414)
2 children	-0.331	-1.328***	-1.358***
	(0.481)	(0.248)	(0.325)
3+ children	-1.456*	-1.133**	-1.388**
	(0.668)	(0.342)	(0.448)
Age of parents			
Wife is aged 25-34	0.885*	-0.134	-0.103
	(0.429)	(0.218)	(0.285)
Wife is aged 45-54	0.187	-0.153	-0.363
	(0.463)	(0.235)	(0.307)
Husband is aged 25-34	0.784	0.413	-0.158
	(0.439)	(0.223)	(0.293)
Husband is aged 45-54	-0.615	0.003	0.287
	(0.433)	(0.220)	(0.288)
Qualifications of parents			
Both partners are university educated	0.452	-0.548**	-1.044***
	(0.437)	(0.222)	(0.291)
Both partners have vocational qualifications	0.901	-0.445	-0.361
	(0.489)	(0.248)	(0.325)
One partner is university educated, other has vocational qualifications	-1.860**	0.051	-0.207
	(0.788)	(0.401)	(0.525)

Table A3   Continued

	Paid Work	*Personal Care*	*Recreation/ Leisure*
One partner is university educated, other has no qualifications	-0.092	-0.425	-0.189
	(0.584)	(0.296)	(0.387)
One partner has vocational qualifications, other has no qualifications	0.683	-0.599**	-0.445
	(0.454)	(0.230)	(0.302)
Labour Force Status			
Both partners work full time		-0.208	-0.558*
		(0.213)	(0.279)
Husband works full time, wife not employed		0.506*	0.497
		(0.220)	(0.288)
Husband not in full time work, other arrangements		0.866***	1.434***
		(0.239)	(0.313)
Household income	0.003***	-0.001***	0.000
	(0.000)	(0.000)	(0.000)
Disabled person in household	-0.972**	-0.103	-0.027
	(0.332)	(0.169)	(0.221)
Hours per week child attends day care	0.072***	0.024**	0.003
	(0.018)	(0.009)	(0.012)
Day of the week			
Both diary days weekend	-8.578***	2.228***	2.792***
	(0.452)	(0.229)	(0.300)
Diary from Saturday and weekday	-5.122***	0.882***	1.737***
	(0.447)	(0.227)	(0.297)
Diary from Sunday and weekday	-4.977***	1.370***	1.413***
	(0.435)	(0.221)	0.289
R square	.398	.195	.180

* P-value<0.05 ** P-value<0.01 *** P-value<0.001. N=1210. Std error in brackets
Source: ABS TUS 1997.

Table A4   **Coefficients of couple parents' joint hours a day in child care sub-categories**

	Physical Care	Interactive Care	Travel/ Commun- ication	Passive Care
Constant term	2.023***	1.090***	0.338**	0.818***
	(0.329)	(0.176)	(0.126)	(0.186)
Number and age of children				
Youngest 0-2				
2 children	0.678**	0.090	0.226**	-0.099
	(0.238)	(0.128)	(0.092)	(0.135)
3+ children	0.281	-0.366**	0.517***	-0.011
	(0.258)	(0.138)	(0.099)	(0.146)
Youngest 3-4				
1 child	-1.327***	-0.077	0.380**	-0.471*
	(0.401)	(0.215)	(0.155)	(0.227)
2 children	-0.572*	-0.362**	0.551***	-0.048
	(0.289)	(0.155)	(0.111)	(0.164)
3+ children	-1.364***	-0.622***	0.592***	-0.058
	(0.322)	(0.173)	(0.124)	(0.182)
Youngest 5-11				
1 child	-2.027***	-0.796***	0.169	-0.638***
	(0.278)	(0.149)	(0.107)	(0.157)
2 children	-1.825***	-0.825***	0.418***	-0.488***
	(0.246)	(0.132)	(0.095)	(0.139)
3+ children	-1.579***	-0.786***	0.582***	-0.520**
	(0.285)	(0.153)	(0.110)	(0.161)
Age of parents				
Wife is aged 25-34	0.149	0.013	-0.092	-0.112
	(0.164)	(0.088)	(0.063)	(0.093)
Wife is aged 45-54	0.048	-0.169	0.047	0.229
	(0.285)	(0.153)	(0.110)	(0.161)
Husband is aged 25-34	0.143	-0.020	0.019	-0.258**
	(0.170)	(0.091)	(0.065)	(0.096)
Husband is aged 45-54	-0.076	0.047	-0.063	-0.209
	(0.222)	(0.119)	(0.085)	(0.125)
Qualifications of parents				
Both partners are university educated	0.464**	0.214*	0.107	0.036
	(0.188)	(0.101)	(0.073)	(0.106)
Both partners have vocational qualifications	0.345	0.143	-0.093	-0.009
	(0.207)	(0.111)	(0.080)	(0.117)
One partner is university educated, other has vocational qualifications	0.699*	0.391*	0.003	0.041
	(0.337)	(0.181)	(0.130)	(0.191)
One partner is university educated, other has no qualifications	0.404*	0.114	-0.032	0.014
	(0.236)	(0.126)	(0.091)	(0.133)

Table A4   Continued

	Physical Care	Interactive Care	Travel/ Commun- ication	Passive Care
One partner has vocational qualifications, other has no qualifications	0.054	0.011	0.056	0.046
	(0.191)	(0.103)	(0.074)	(0.108)
Labour Force Status				
Both partners work full time	0.052	-0.066	-0.239**	-0.161
	(0.192)	(0.103)	(0.074)	(0.109)
Husband works full time, wife not employed	0.302	0.060	0.070	-0.043
	(0.167)	(0.089)	(0.064)	(0.094)
Husband not in full time work, other arrangements	0.199	0.254**	-0.108	-0.006
	(0.191)	(0.102)	(0.073)	(0.108)
Household income	0.000	0.000	0.000	0.000
	(0.000)	(0.000)	(0.000)	(0.000)
Disabled person in household	-0.008	-0.104	0.010	0.070
	(0.144)	(0.077)	(0.055)	(0.081)
Hours per week child attends day care	-0.022**	-0.004	0.006**	-0.005
	(0.006)	(0.003)	(0.002)	(0.004)
Day of the week				
Both diary days weekend	0.053	0.304**	-0.466***	0.161
	(0.182)	(0.098)	(0.070)	(0.103)
Diary from Saturday and weekday	-0.052	0.103	-0.205**	0.249*
	(0.194)	(0.104)	(0.075)	(0.110)
Diary from Sunday and weekday	-0.302	0.163	-0.190**	0.121
	(0.177)	(0.095)	(0.068)	(0.100)
R square	.329	.195	.191	.080

* P-value<0.05   ** P-value<0.01   *** P-value<0.001   N=705 households   Std error in brackets

Source: ABS TUS 1997.

Table A5   **Coefficients of couple parents' joint hours a day in child care as a primary activity, child care as either a primary or a secondary activity, paid work and domestic labour**

	Childcare (primary)	Childcare (primary and secondary)	Paid work	Domestic Labour
Constant term	4.403***	8.523***	6.752***	4.864***
	(0.435)	(0.915)	(0.846)	(0.497)
Number and age of children				
Youngest 0-2				
2 children	0.906**	1.221	0.610	0.533
	(0.322)	(0.677)	(0.688)	(0.360)
3+ children	0.430	0.361	-0.496	
	(0.348)	(0.732)	(0.742)	0.759
Youngest 3-4				(0.390)
1 child	-1.569**	-2.180	1.424	1.542**
	(0.543)	(1.143)	(1.166)	(0.607)
2 children	-0.429	0.224	0.839	0.480
	(0.391)	(0.823)	(0.836)	(0.438)
3+ children	-1.491**	-0.649	0.812	0.907
	(0.436)	(0.917)	(0.929)	(0.488)
Youngest 5-11				
1 child	-3.239***	-5.379***	1.927**	0.952**
	(0.377)	(0.793)	(0.806)	(0.421)
2 children	-2.665***	-4.245***	2.972***	0.001
	(0.332)	0.697	(0.708)	(0.372)
3+ children	-2.128***	-4.276***	1.807	1.073*
	(0.385)	(0.809)	(0.825)	(0.431)
Age of parents				
Wife is aged 25-34	-0.039	0.012	-0.112	-0.430
	(0.222)	(0.466)	(0.474)	(0.248)
Wife is aged 45-54	0.130	-0.240	-0.506	-0.118
	(0.386)	(0.812)	(0.820)	(0.431)
Husband is aged 25-34	-0.163	0.177	0.925	-0.091
	(0.230)	(0.483)	(0.490)	(0.257)
Husband is aged 45-54	-0.315	-0.381	-0.659	0.018
	(0.300)	(0.631)	(0.643)	(0.335)
Qualifications of parents				
Both partners are university educated	0.835**	2.486***	0.142	-0.546
	(0.249)	(0.523)	(0.534)	(0.285)
One partner is university educated, other has vocational qualifications or no qualifications	0.513*	1.528**	-0.463	0.646
	(0.262)	(0.552)	(0.562)	(0.510)
Both partners have vocational qualifications	0.410	1.454**	0.027	-0.336
	(0.241)	(0.506)	(0.516)	(0.313)

Table A5   Continued

	Childcare (primary)	Childcare (primary and secondary)	Paid work	Domestic Labour
One partner has vocational qualifications, other has no qualifications	0.155	0.673	0.225	0.054
	(0.254)	(0.533)	(0.543)	(0.289)
Labour Force Status				
Both partners work full time	-0.449	-1.173*		-0.835**
	(0.261)	(0.548)		(0.291)
Husband works full time, wife not employed	0.428	-0.106		0.295
	(0.225)	(0.473)		(0.252)
Husband not in full time work, other arrangements	0.392	0.709		0.461
	(0.258)	(0.543)		(0.288)
Household income	0.000	0.001	0.002***	0.000
	(0.000)	(0.000)	(0.000)	(0.000)
Hours per week child attends day care	-0.023*	-0.075***	0.080***	0.515
	(0.009)	(0.018)	(0.017)	(0.218)
Disabled person in household	-0.047	-0.205	-0.798	-0.014
	(0.195)	(0.410)	(0.417)	(0.010)
Day of the week				
Both diary days weekend	-0.013	1.491**	-7.700***	1.561***
	(0.246)	(0.518)	(0.528)	(0.275)
Diary from Saturday and weekday	0.006	1.156*	-4.889***	0.834**
	(0.262)	(0.552)	(0.563)	(.294)
Diary from Sunday and weekday	-0.205	1.170*	-4.436***	1.249***
	(0.239)	(0.502)	(0.512)	(0.269)
R square	.342	.274	.377	.185

* P-value<0.05. ** P-value<0.01. *** P-value<0.001. N=705 households. Std error in bracket

Source: ABS TUS 1997.

Table A6 **Coefficients of hours a day spent by men and women in couple-headed households in child care, unpaid work and total work as a primary activity and total work as a primary or secondary activity**

	Child care		Unpaid Work		Total Work		Total Work (primary and secondary activity)	
	Male	Female	Male	Female	Male	Female	Male	Female
Constant term	-0.274**	-0.249	0.863***	3.742***	7.504***	7.447***	7.362***	6.964***
	(0.109)		(0.243)	(0.277)	(0.316)	(0.275)	(0.340)	(0.364)
Number and age of children								
Youngest 0-2								
1 child	1.404***	3.358***	1.545***	3.571***	0.839*	1.236***	2.411***	5.507***
	(0.107)	(0.151)	(0.239)	(0.267)	(0.333)	(0.282)	(0.358)	(0.372)
2 children	1.461***	3.964***	1.920***	4.464***	2.118***	2.330***	3.988***	6.906***
	(0.108)	(0.151)	(0.242)	(0.266)	(0.335)	(0.273)	(0.361)	(0.362)
3+ children	1.259***	3.883***	1.513***	4.816***	1.312***	1.824***	3.184***	6.077***
	(0.115)	(0.158)	(0.256)	(0.280)	(0.355)	(0.287)	(0.383)	(0.379)
Youngest 3-4								
1 child	1.038***	2.134***	1.907***	3.288***	2.023**	1.352**	3.337***	5.452***
	(0.192)	(0.266)	(0.428)	(0.469)	(0.595)	(0.507)	(0.641)	(0.671)
2 children	1.403***	2.932***	1.375***	3.686***	1.305**	1.460***	2.698***	6.243***
	(0.132)	(0.188)	(0.295)	(0.331)	(0.410)	(0.347)	(0.441)	(0.460)
3+ children	0.825***	2.377***	0.912**	3.257***	1.202**	1.487***	2.981***	6.441***
	(0.144)	(0.203)	(0.322)	(0.358)	(0.447)	(0.377)	(0.482)	(0.499)
Youngest 5-11								
1 child	0.448***	1.052***	1.138***	1.700***	0.881**	0.626*	1.864***	3.097***
	(0.102)	(0.142)	(0.227)	(0.251)	(0.316)	(0.272)	(0.340)	(0.360)
2 children	0.617***	1.423***	0.571**	2.034***	1.484***	0.985***	2.867***	3.708***
	(0.084)	(0.122)	(0.187)	(0.215)	(0.259)	(0.227)	(0.279)	(0.300)
3+ children	0.736***	1.658***	0.971***	2.771***	1.092***	1.134***	2.114***	3.591***
	(0.112)	(0.158)	(0.249)	(0.278)	(0.346)	(0.298)	(0.373)	(0.394)
Age								
25-34	-0.125*	0.051	-0.169	-0.276	0.288	-0.021	-0.053	0.107
	(0.063)	(0.085)	(0.140)	(0.150)	(0.194)	(0.165)	(0.209)	(0.218)
45-54	-0.071	-0.042	0.508**	0.433	-0.506**	-0.163	-0.640**	0.176
	(0.071)	(0.116)	(0.159)	(0.205)	(0.220)	(0.223)	(0.237)	(0.296)
Qualifications								
Postgraduate	0.283**	0.353**	0.345	-0.362	-0.190	0.676*	0.457	1.037**
	(0.112)	(0.162)	(0.251)	(0.286)	(0.348)	(0.313)	(0.375)	(0.414)
Bachelor degree	0.248**	0.233*	0.513**	0.075	0.358	0.375	1.037***	0.761**
	(0.089)	(0.117)	(0.197)	(0.207)	(0.274)	(0.227)	(0.296)	(0.300)
University diploma	0.107	0.363**	0.194	0.313	0.456	0.684**	0.714**	1.431***
	(0.085)	(0.116)	(0.190)	(0.205)	(0.264)	(0.224)	(0.284)	(0.297)
Skilled vocational	0.053	0.238**	0.174	0.151	0.742***	0.155	0.821***	0.398
	(0.061)	(0.106)	(0.136)	(0.187)	(0.188)	(0.205)	(0.202)	(0.271)
Basic vocational	-0.043	-0.007	-0.168	0.269	0.993	0.508	1.466**	0.661
	(0.176)	(0.148)	(0.393)	(0.262)	(0.541)	(0.286)	(0.583)	(0.379)
Usual hours worked by spouse/partner	0.003	0.005	0.006	0.012**	0.011*	0.010*	0.020**	0.018**
	(0.002)	(0.002)	(0.004)	(0.004)	(0.005)	(0.005)	(0.006)	(0.006)
Labour Force Status								
Full time		-0.181		-1.358***				
		(0.095)		(0.168)				
Part time	0.381**		0.709**					
	(0.119)		(0.264)					
Not in the labour force	0.461***	0.478	1.491***	0.944***				
	(0.094)	(0.089)	(0.210)	(0.157)				
Household income	0.000	0.000	0.000	0.000	0.001***	0.001***	0.001***	0.001**
	(0.000)	(0.000)	(0.000)	(0.000)	(0.000)	(0.000)	(0.000)	(0.000)
Hours per week child attends day care	-0.004	-0.024***	-0.006	-0.035***	-0.009	0.006	-0.017	-0.042***
	(0.003)	(0.004)	(0.006)	(0.007)	(0.009)	(0.007)	(0.009)	(0.010)
Disabled person in household	0.017	-0.015	0.357**	0.052	0.084	-0.307*	0.128	-0.468*
	(0.057)	(0.078)	(0.127)	(0.138)	(0.177)	(0.150)	(0.190)	(0.199)
Day of the week								
Saturday	0.126	-0.170	1.722***	0.493**	-3.315***	-2.159***	-2.536***	-1.664***
	(0.074)	(0.102)	(0.165)	(0.180)	(0.229)	(0.197)	(0.247)	(0.261)
Sunday	0.334***	-0.310	1.891***	-0.168	-4.009***	-2.616***	-3.084***	-2.101***
	(0.072)	(0.099)	(0.161)	(0.174)	(0.223)	(0.191)	(0.240)	(0.252)
R square	.213	.492	.188	.375	.248	.197	.228	3.17

* P-value <0.05   ** P-value<0.01   *** P-value<0.001. N=4274 person-days. Std error in brackets

Source: ABS TUS 1997.

Table A7 **Coefficients of hours a day spent by men and women in couple-headed households in employment, personal care, recreation and domestic labour as a primary activity**

	Paid Work		Personal Care		Recreation/Leisure		Domestic Labour	
	Male	Female	Male	Female	Male	Female	Male	Female
Constant term	5.831***	3.503***	10.892***	11.589***	3.532***	3.565***	1.137***	3.992***
	(0.389)	(0.309)	(0.200)	(0.187)	(0.249)	(0.217)	(0.207)	(0.238)
Number and age of children								
Youngest 0-2								
1 child	-0.980**	-3.067***	-0.450*	-0.610**	-0.084	-0.978***	0.141	0.212
	(0.410)	(0.317)	(0.198)	(0.180)	(0.245)	(0.209)	(0.204)	(0.229)
2 children	0.313	-3.183***	-0.607**	-1.350***	-1.006***	-1.420***	0.459*	0.499*
	(0.413)	(0.308)	(.199)	(0.180)	(0.247)	(0.208)	(0.206)	(0.228)
3+ children	-0.515	-4.067***	-0.783***	-1.739***	-0.354	-0.745**	0.253	0.933***
	(0.438)	(0.323)	(0.211)	(0.189)	(0.262)	(0.219)	(0.218)	(0.240)
Youngest 3-4								
1 child	0.161	-2.745***	-0.853**	-1.262***	-0.700	-0.395	0.869*	1.154**
	(0.733)	(0.571)	(0.353)	(0.317)	(0.438)	(0.368)	(0.365)	(0.403)
2 children	-0.347	-3.247***	-0.530*	-1.120***	-0.738**	-1.191***	-0.028	0.754**
	(0.505)	(0.391)	(0.243)	(0.224)	(0.302)	(0.260)	(0.251)	(0.284)
3+ children	0.067	-2.855***	-0.748**	-1.475***	-0.457	-0.886**	0.087	0.880**
	(0.551)	(0.425)	(0.266)	(0.242)	(0.329)	(0.281)	(0.275)	(0.307)
Youngest 5-11								
1 child	-0.302	-1.606***	-0.447*	-0.654***	-0.358	-0.394*	0.690***	0.648**
	(0.389)	(0.307)	(0.188)	(0.170)	(0.233)	(0.197)	(0.194)	(0.216)
2 children	0.864**	-1.778***	-0.766***	-0.878***	-0.454**	-0.632***	-0.046	0.612**
	(0.319)	(0.255)	(0.154)	(0.145)	(0.191)	(0.168)	(0.159)	(0.185)
3+ children	0.154	-2.446***	-0.550**	-0.828***	-0.298	-0.762***	0.235	1.112
	(0.427)	(0.335)	(0.206)	(0.188)	(0.255)	(0.218)	(0.212)	(0.239)
Age								
25-34	0.481*	0.212	0.113	-0.019	-0.077	0.100	-0.044	-0.326*
	(0.239)	(0.185)	(0.116)	(0.102)	(0.143)	(0.118)	(0.119)	(0.129)
45-54	-1.079***	-0.803**	0.132	-0.226	0.215	0.111	0.580***	0.476**
	(0.271)	(0.252)	(0.131)	(0.139)	(0.162)	(0.161)	(0.135)	(0.176)
Qualifications								
Postgraduate	-0.940	0.997**	0.041	-0.425*	-0.204	-0.235	0.062	-0.715**
	(0.429)	(0.353)	(0.207)	(0.193)	(0.256)	(0.224)	(0.214)	(0.246)
Bachelor degree	-0.350	0.270	-0.173	-0.336	-0.316	0.062	0.265	-0.158
	(0.338)	(0.255)	(0.163)	(0.140)	(0.202)	(0.162)	(0.168)	(0.178)
University diploma	0.282	0.221	-0.226	-0.270	-0.173	-0.337*	0.087	-0.050
	(0.325)	(0.252)	(0.157)	(0.138)	(0.194)	(0.160)	(0.162)	(0.176)
Skilled vocational qualifications	0.716**	0.059	-0.333**	0.035	-0.291*	-0.008	0.121	-0.087
	(0.231)	(0.231)	(0.112)	(0.126)	(0.139)	(0.147)	(0.116)	(0.161)
Basic vocational	1.217	0.392	-0.358	-0.290	-0.198	0.017	-0.125	0.276
	(0.667)	(0.322)	(0.324)	(0.177)	(0.402)	(0.205)	(0.335)	(0.225)
Usual hours worked by spouse/partner	0.001	0.000	-0.007*	-0.003	-0.002	-0.004	0.003	0.007
	(0.007)	(0.005)	(0.003)	(0.003)	(0.004)	(0.003)	(0.003)	(0.004)
Labour Force Status								
Full time				-0.432***		-0.398***		-1.177***
				(0.113)		(0.132)		(0.144)
Part time			0.025		0.266		0.329	
			(0.218)		(0.271)		(0.225)	
Not in the labour force			1.017***	0.290**	2.141***	0.642***	1.030	0.466**
			(0.173)	(0.106)	(0.215)	(0.123)	(0.179)	(0.135)
Household income	0.002***	0.001***	0.000**	0.000	0.000	0.000	0.000	0.000
	(0)	(0)	(0)	(0)	(0)	(0)	(0)	(0)
Hours per week child attends day care	0.001	0.065***	0.012*	0.013**	-0.006	0.005	-0.002	-0.012*
	(0.011)	(0.008)	(0.005)	(0.005)	(0.006)	(0.005)	(0.005)	(0.006)
Disabled person in household	-0.291	-0.585**	-0.122	0.080	0.007	-0.058	0.340**	0.067
	(0.218)	(0.169)	(0.105)	(0.093)	(0.130)	(0.108)	(0.109)	(0.118)
Day of the week								
Saturday	-4.920***	-2.589***	0.751***	0.794***	1.668***	0.786***	1.596***	0.662***
	(0.282)	(0.222)	(0.136)	(0.121)	(0.169)	(0.141)	(0.141)	(0.154)
Sunday	-5.867***	-2.417***	1.577***	1.139***	1.620***	0.946***	1.557***	0.142
	(0.275)	(0.215)	(0.133)	(0.117)	(0.165)	(0.136)	(0.137)	(0.149)
R square	.305	.258	.131	.130	.164	.093	.160	.157

* P-value<0.05. ** P-value<0.01. *** P-value<0.001. N=4274 person-days. Std error in brackets

Source: ABS TUS 1997.

Table A8 **Coefficients of minutes a day spent by mothers and fathers in couple-headed households in child care sub-categories as a primary activity**

	Physical Care		Interactive Care		Travel/ Communication		Passive Care	
	Male	Female	Male	Female	Male	Female	Male	Female
Constant term	12.758**	111.305***	25.024***	56.076***	0.571	8.005	22.537***	31.688***
	(4.442)	(11.494)	(4.513)	(6.056)	(2.755)	(5.407)	(4.453)	(6.639)
Number and age of children								
Youngest 0-2								
2 children	7.956*	22.263**	4.875	3.880	2.431	10.422**	-4.243	-5.271
	(3.838)	(8.373)	(3.899)	(4.412)	(2.380)	(3.939)	(3.847)	(4.836)
3+ children	4.665	6.435	-8.641*	-9.984*	6.371**	26.446***	-7.074	5.619
	(4.172)	(9.160)	(4.239)	(4.826)	(2.588)	(4.309)	(4.182)	(5.291)
Youngest 3-4								
1 child	-8.792	-66.706***	-4.513	-0.145	4.744	16.028**	-14.133*	-16.917*
	(6.486)	(14.112)	(6.590)	7.436	(4.023)	(6.638)	(6.502)	(8.151)
2 children	4.433	-42.581***	-1.201	-14.615**	6.746**	25.982***	-8.763	5.273
	(4.668)	(10.268)	(4.743)	(5.410)	(2.895)	(4.830)	(4.679)	(5.930)
3+ children	-8.436	-79.716***	-21.118***	-9.073	7.513**	29.984***	-11.410*	4.969
	(5.191)	(11.322)	(5.274)	(5.966)	(3.220)	(5.326)	(5.204)	(6.539)
Youngest 5-11								
1 child	-20.639***	-108.032***	-16.795***	-20.413***	3.002	8.512	-20.502***	-22.159***
	(4.414)	(9.826)	(4.484)	(5.177)	(2.737)	(4.622)	(4.424)	(5.675)
2 children	-17.260	-98.253***	-18.576***	-22.680***	5.261*	20.390***	-13.598***	-17.659***
	(3.847)	(8.390)	(3.909)	(4.421)	(2.386)	(3.947)	(3.856)	(4.846)
3+ children	-13.407**	-87.932***	-15.473**	-19.024***	6.756**	25.113***	-17.165***	-17.920**
	(4.450)	(9.761)	(4.521)	(5.143)	(2.760)	(4.592)	(4.461)	(5.638)
Age								
25-34	1.205	16.837**	-0.955	-4.224	-0.454	-2.963	-8.567**	-11.574***
	(2.550)	(5.368)	(2.591)	(2.828)	(1.581)	(2.525)	(2.556)	(3.100)
45-54	-0.607	-3.172	-3.261	-6.747	0.623	-0.569	-2.173	6.021
	(3.083)	(9.578)	(3.132)	(5.047)	(1.912)	(4.506)	(3.091)	(5.532)
Qualifications								
Postgraduate	2.942	26.884**	9.406*	11.973*	3.161	1.329	6.027	-1.209
	(4.353)	(10.107)	(4.422)	(5.325)	(2.699)	(4.754)	(4.363)	(5.837)
Bachelor degree	12.262***	13.808*	9.845**	5.713	1.966	4.215	0.202	-3.106
	(3.469)	(8.199)	(3.524)	(4.320)	(2.151)	(3.857)	(3.477)	(4.736)
University diploma	4.163	15.582*	6.830*	10.872**	-0.993	4.046	0.357	7.454
	(3.424)	(7.112)	(3.479)	(3.748)	(2.124)	(3.346)	(3.432)	(4.108)
Skilled vocational	1.048	25.735***	2.762	2.823	1.198	-5.416	1.906	-1.285
	(2.478)	(6.615)	(2.517)	(3.485)	(1.536)	(3.112)	(2.483)	(3.820)
Basic vocational	9.147	13.845	-1.779	6.373	-5.116	-12.979**	-1.676	-5.776
	(7.245)	(9.655)	(7.360)	(5.087)	(4.493)	(4.542)	(7.262)	(5.576)
Usual hours worked by spouse/partner	0.135	0.154	-0.004	-0.165*	0.093*	0.299***	0.037	-0.056
	(0.070)	(0.153)	(0.071)	(0.081)	(0.043)	(0.072)	(0.070)	(0.089)
Labour Force Status								
Full time		10.332		-5.630	1.666	-9.489**		-4.176
		(6.751)		(3.557)	(2.992)	(3.176)		(3.899)
Part time	12.943	19.401**	8.445		4.070		10.433*	
	(4.825)	(5.624)	(4.902)	1.427	(2.411)		(4.836)	-0.180
Not in the labour force	8.911		18.850***			1.890	3.387	
	(3.887)		(3.949)	(2.963)		(2.646)	(3.897)	(3.248)
Household income	0.005	-0.023*	0.002	-0.009	-0.001	-0.008	-0.004	-0.010
	(0.003)	(0.011)	(0.006)	(0.006)	(0.002)	(0.005)	(0.003)	(0.006)
Hours per week child attends day care	-0.030	-1.267***	-0.103	-0.130	0.108	0.208*	-0.210*	-0.041
	(0.099)	(0.217)	(0.101)	(0.114)	(0.062)	(0.102)	(0.100)	(0.125)
Disabled person in household	3.145	-1.441	-3.547	-3.261	2.243	-0.950	-1.767	5.894*
	(2.324)	(5.084)	(2.361)	(2.679)	(1.441)	(2.391)	(2.330)	(2.936)
Day of the week								
Saturday	4.326	-6.708	4.436	4.625	-1.127	-17.899	5.206	7.998*
	(2.999)	(6.524)	(.047)	(3.437)	(1.860)	(3.069)	(3.006)	(3.768)
Sunday	6.294*	-6.555	13.436***	-1.158	-1.068	-23.665***	7.357**	8.178**
	(2.833)	(6.158)	(2.878)	(3.245)	(1.757)	(2.897)***	(2.839)	(3.557)
R square	0.95	.323	.091	.069	.026	.157	.038	.056

* P-value<0.05   ** P-value<0.01   *** P-value<0.001   N=2926 person-days   Std error in brackets

Source: ABS TUS 1997.

Table A9   **Coefficients of minutes a day spent by mothers and fathers in couple-headed households in child care sub-categories as primary or secondary activity**

	Physical care		Interactive care		Travel/communication		Passive care	
	Male	*Female*	*Male*	*Female*	*Male*	*Female*	*Male*	*Female*
Constant	15.175**	115.576***	37.688***	92.688***	1.027	8.965	58.932**	241.258***
	(4.817)	(12.082)	(7.847)	(14.687)	(2.836)	(5.679)	(17.041)	(30.945)
Number and age of children								
Youngest 0-2								
2 children	6.511	24.229**	13.879*	35.850**	3.365	11.478**	-3.967	-17.604
	(4.162)	(8.801)	(6.780)	(10.699)	(2.450)	(4.137)	(14.724)	(22.543)
3+ children	3.707	14.925	-4.177	30.432**	6.761**	29.165***	-7.076	-41.858
	(4.525)	(9.628)	(7.371)	(11.704)	(2.664)	(4.526)	(16.007)	(24.661)
Youngest 3-4								
1 child	-12.739	-73.372***	5.348	7.774	4.719	16.282**	-59.720**	-49.674
	(7.034)	(14.834)	(11.458)	(18.032)	(4.141)	(6.972)	(24.883)	(37.993)
2 children	2.065	-44.525***	-8.343	30.532*	7.044**	26.522***	-13.838	9.920
	(5.062)	(10.793)	(8.246)	(13.120)	(2.980)	(5.073)	(17.908)	(27.643)
3+ children	-12.217*	-84.653***	-16.517	38.574**	7.608**	30.585***	-16.600	-17.396
	(5.629)	(11.901)	(9.171)	(14.467)	(3.314)	(5.594)	(19.916)	(30.482)
Youngest 5-11								
1 child	-23.891***	-116.588***	-21.310**	-16.279	2.427	7.853	-72.32***	-157.185***
	(4.786)	(10.328)	(7.797)	(12.555)	(2.818)	(4.855)	(16.932)	(26.453)
2 children	-21.067***	-104.696***	-14.963*	-5.314	4.961*	20.979***	-42.915**	-156.416***
	(4.172)	(8.819)	(6.796)	(10.721)	(2.456)	(4.145)	(14.760)	(22.588)
3+ children	-17.392***	-90.717***	-13.044	4.102	7.732**	25.293***	-69.765	-170.310***
	(4.826)	(10.260)	(7.861)	(12.472)	(2.841)	(4.823)	(17.072)	(26.280)
Age								
25-34	1.544	16.938**	-2.223	-14.831*	-0.791	-3.657	-20.145*	27.340
	(2.765)	(5.642)	(4.505)	(6.858)	(1.628)	(2.652)	(9.783)	(14.451)
45-54	-0.997	-2.784	-4.946	-27.617*	1.022	-1.197	-13.848	24.418
	(3.343)	(10.068)	(5.447)	(12.239)	(1.969)	(4.732)	(11.828)	(25.787)
Qualifications								
Postgraduate	5.599	29.441**	18.563**	33.699***	2.787	0.862	53.447**	42.842
	(4.720)	(10.623)	(7.689)	(12.914)	(2.779)	(4.993)	(16.697)	(27.209)
Bachelor degree	13.779***	15.669*	18.143**	19.641*	2.116	3.689	46.456***	60.135**
	(3.761)	(8.619)	(6.127)	(10.477)	(2.215)	(4.051)	(13.307)	(22.075)
University diploma	5.050	22.814**	9.208	11.061	(-0.773)	3.908	26.660*	64.057**
	(3.713)	(7.476)	(6.049)	(9.088)	(2.186)	(3.514)	(13.137)	(19.148)
Skilled vocational	1.619	25.252***	3.677	6.524	1.055	-4.180	11.323	40.958
	(2.687)	(6.953)	(4.377)	(8.452)	(1.582)	(3.268)	(9.505)	(17.808)
Basic vocational	12.833	16.049	7.710	-4.974	-5.185	-13.901**	33.582	45.646
	(7.856)	(10.149)	(12.798)	(12.337)	(4.625)	(4.770)	(27.793)	(25.993)
Usual hours worked by spouse/partner	0.141	0.288	0.063	0.410*	0.119**	0.279***	0.814**	0.329
	(0.075)	(0.161)	(0.123)	(0.196)	(0.044)	(0.076)	(0.267)	(0.413)
Labour Force Status								
Full time		9.822		-15.632		-8.370**		-31.560
		(7.096)		(8.626)		(3.335)		(18.174)
Part time	15.979**		12.808		1.002		51.465**	
	(5.232)		(8.524)		(3.081)		(18.510)	
Not in the labour force	10.078**	20.712***	20.924**	1.789	4.123	2.194	62.562***	2.708
	(4.215)	(5.912)	(6.867)	(7.186)	(2.482)	(2.779)	(14.913)	(15.142)
Household income	0.006	-0.021	0.004	-0.013	-0.002	-0.007	0.016	-0.062*
	(0.003)	(0.012)	(0.006)	(0.014)	(0.002)	(0.006)	(0.012)	(0.030)
Hours per week child attends day care	-0.032	-1.374	-0.214	-0.973	0.084	0.162	-0.854*	-2.125***
	(0.108)	(0.228)	(0.176)	(0.277)	(0.063)	(0.107)	(0.381)	(0.584)
Disabled person in household	3.385	-0.826	-0.297	-6.350	2.327	-1.379	3.082	-11.337
	(2.520)	(5.343)	(4.105)	(6.495)	(1.484)	(2.512)	(8.916)	(13.686)
Day of the week								
Saturday	4.907	-7.852	15.878**	-2.451	-0.252	-16.758***	47.839	61.373***
	(3.252)	(6.857)	(5.298)	(8.335)	(1.915)	(.223)	(17.507)	(17.563)
Sunday	7.089**	-9.310	28.285***	-14.210	-1.295	-24.178***	66.481***	65.994***
	(3.072)	(6.473)	(5.004)	(7.868)	(1.808)	(3.042)	(10.866)	(16.579)
R square	.102	.333	.066	0.66	.028	.137	.088	.143

* P-value<0.05. ** P-value<0.01. *** P-value<0.001. N=2926 person-days. Std error in brackets

Source: ABS TUS 1997.

Table A10 **Coefficients of hours a day spent by mothers and fathers in couple-headed or single-parent households in child care, unpaid work, total work as a primary activity and total work as either a primary or a secondary activity**

	Childcare (primary)	Unpaid work (primary)	Total work (primary)	Total Work (primary and secondary)
Constant term	0.172***	2.875***	7.193	7.477***
	(0.077)	(0.156)	0.197	(0.230)
Family configuration				
Single				
Man with no children	0.042	-0.349*	0.120	0.104
	(0.080)	(0.162)	(0.204)	(0.238)
Mother with youngest child under 5	2.801***	4.580***	2.047***	5.816***
	(0.164)	(0.332)	(0.418)	(0.488)
Mother with youngest child aged 5-11	1.326***	2.920***	0.513	3.413***
	(0.132)	(0.266)	(0.335)	(0.391)
Partnered				
Woman with no children	-0.013	0.982***	0.758***	0.780**
	(0.076)	(0.153)	(0.192)	(0.224)
Man with no children	0.003	-0.592***	0.659***	0.521*
	(0.078)	(0.158)	(0.199)	(0.233)
Mother with youngest child under 5	3.002***	4.760***	2.576***	6.190***
	(0.092)	(0.186)	(0.234)	(0.273)
Father with youngest child under 5	0.934***	0.390*	1.664***	2.735***
	(0.092)	(0.185)	(0.233)	(0.272)
Mother with youngest child aged 5-11	1.119***	2.874***	1.660***	3.621***
	(0.093)	(0.187)	(0.236)	(0.275)
Father with youngest child aged 5-11	0.334***	-0.352	1.452***	2.220***
	(0.094)	(0.190)	(0.240)	(0.280)
Age				
25-34	-0.024	-0.377***	-0.119	-0.223
	(0.041)	(0.084)	(0.105)	(0.123)
45-54	-0.058	0.603***	-0.262*	-0.289
	(0.051)	(0.102)	(0.129)	(0.150)
Qualification				
Postgraduate	0.296***	0.162	0.292	0.768**
	(0.077)	(0.155)	(0.195)	(0.227)
Bachelor degree	0.212***	0.177	0.340**	0.815***
	(0.058)	(0.117)	(0.147)	(0.172)
University diploma	0.256***	0.167	0.632***	1.146***
	(0.057)	(0.116)	(0.146)	(0.170)
Skilled vocational qualifications	0.101*	-0.016	0.747***	0.843***
	(0.045)	(0.091)	(0.115)	(0.134)
Basic vocational qualifications	0.085	0.196***	0.569**	0.790**
	(0.078)	(0.157)	(0.198)	(0.231)
Household weekly income	0.000	-0.001***	0.002***	0.001***
	(0.000)	(0.000)	(0.000)	(0.000)
Disabled person in household	0.040	0.240**	-0.369***	-0.460***
	(0.038)	(0.077)	(0.097)	(0.113)
Number of children under 15 years in household	0.139***	0.253***	0.079	0.221**
	(0.026)	(0.052)	(0.065)	(0.076)
Day of the week				
Saturday	-0.039	1.041***	-2.727***	-2.234***
	(.050)	(0.101)	(0.128)	(0.149)
Sunday	-0.003	0.805***	-3.304***	-2.756***
	(0.050)	(0.100)	(0.127)	(0.148)
R square	.455	.413	.217	.300

* P-value<0.05. ** P-value<0.01. *** P-value<0.001. N=6035 person-days. Std error in brackets

Source: ABS TUS 1997.

Table A11   Coefficients of hours a day spent by mothers in couple-headed or single-parent households in child care as either a primary or a secondary activity, child-free recreation, time in the company of children, and time in sole charge of children

	Childcare (primary and secondary)	Child-free recreation	Time with children	Time alone with children
Constant term	6.012***	0.097	12.320***	3.894***
	(0.476)	(0.122)	(0.466)	(0.494)
Single mother	1.379**	0.324**	0.745	5.734***
	(0.397)	(0.102)	(0.389)	(0.412)
Age				
25-34	0.533**	0.047	0.314	0.935***
	(0.226)	(0.058)	(0.222)	(0.235)
45-54	0.331	-0.118	0.178	-1.312**
	(0.412)	(0.106)	(0.404)	(0.428)
Qualification				
Postgraduate	0.944*	-0.193	0.365	-0.209
	(0.462)	(0.118)	(0.453)	(0.480)
Bachelor degree	0.906**	-0.030	-0.013	0.613
	(0.355)	(0.091)	(0.348)	(0.368)
University diploma	1.755***	0.033	0.221	0.437
	(0.317)	(0.081)	(0.311)	(0.329)
Skilled vocational qualifications	0.767**	0.021	0.036	-0.279
	(0.284)	(0.073)	(0.278)	(0.295)
Basic vocational qualifications	0.253	0.069	0.163	0.614
	(0.389)	(0.100)	(0.381)	(0.403)
Hours worked by spouse	0.020**	0.000	0.032***	0.065***
	(0.008)	(0.002)	(0.007)	(0.008)
Household weekly income	0.000	0.000*	-0.001***	-0.001*
	(0.000)	(0.000)	(0.000)	(0.000)
Hours child usually attends day care	-0.064***	0.000	-0.090***	-0.045***
	(0.008)	(0.002)	(0.008)	(0.009)
Disabled person in household	-0.421	0.120*	0.076	-0.306
	(0.218)	(0.056)	(0.214)	(0.226)
Youngest child is aged 5-11	-3.616***	0.389	-3.199***	-2.747***
	(0.234)	(0.060)	(0.230)	(0.243)
Number of children under 15 years in household	0.363**	-0.032	0.319**	-0.210
	(0.113)	(0.029)	(0.111)	(0.118)
Day of the week				
Saturday	0.325	0.161*	1.795***	-1.873***
	(0.284)	(0.073)	(0.278)	(0.294)
Sunday	(0.165)	-0.094	1.942***	-2.368***
	0.271	(0.069)	(0.266)	(0.281)
R square	.202	.061	.229	.266

* P-value<0.05. ** P-value<0.01. *** P-value<0.001. N=1708. Std error in brackets
Source: ABS TUS 1997.

Table A12   Coefficients of minutes a day spent by mothers in couple-headed or single-parent households in child care sub-categories (primary and secondary activity)

	Physical Care	*Interactive Care*	*Travel/ Commun- ication*	*Passive Care*
Constant term	95.166***	81.682***	4.986	205.890***
	(10.487)	(13.133)	(5.014)	(28.481)
Single mother	3.254	21.556*	13.331**	53.392*
	(8.746)	(10.952)	(4.181)	(23.751)
Age				
25-34	23.796***	-8.024	-1.812	26.078
	(4.985)	(6.242)	(2.383)	(13.538)
45-54	2.225	-14.652	-3.377	23.494
	(9.091)	(11.384)	(4.346)	(24.688)
Qualification				
Postgraduate	22.891*	24.696	-2.609	25.396
	(10.190)	(12.761)	(4.872)	(27.674)
Bachelor degree	9.234	18.705	3.304	33.594
	(7.822)	(9.795)	(3.739)	(21.241)
University diploma	15.837*	4.786	1.280	89.597***
	(6.987)	(8.750)	(3.341)	(18.976)
Skilled vocational qualifications	22.759***	(2.490)	-6.091*	34.912*
	(6.263)	(7.843)	(2.994)	(17.009)
Basic vocational qualifications	16.270	(-6.969)	-11.974***	26.171
	(8.569)	(10.731)	(4.097)	(23.271)
Hours worked by spouse	0.259	(0.359)	0.356***	0.524
	(0.166)	(0.208)	(0.079)	(0.451)
Household weekly income	-0.009	(0.009)	(-0.003)	-0.018
	(0.006)	(0.007)	(0.003)	(0.016)
Hours child usually attends day care	-1.320***	-1.230***	0.090	-1.819***
	(0.184)	(0.231)	(0.088)	(0.500)
Disabled person in household	-0.853	-13.094*	1.214	-16.702
	4.808	(6.020)	(2.299)	(13.056)
Youngest child is aged 5-11	(-86.781)***	-28.968***	2.874	-136.934***
	5.170	(6.475)	(2.472)	(14.042)
Number of children under 15 years in household	4.798	11.357***	7.939***	1.423
	(2.499)	(3.129)	(1.195)	(6.786)
Day of the week				
Saturday	-7.094	-9.198	-17.688***	55.693**
	(6.255)	(7.834)	(2.991)	(16.988)
Sunday	-5.121	-16.562*	-24.031***	62.491***
	(5.977)	(7.485)	(2.858)	(16.233)
R square	.251	.052	.100	.107

* P-value<0.05. ** P-value<0.01. *** P-value<0.001. N=1708. Std error in brackets
Source: ABS TUS 1997.

Table A13 Coefficients of hours a day spent by mothers and fathers in couple-headed households in child care as either a primary or a secondary activity, in the company of children, in paid work and in domestic labour (shorter model)

	Childcare (primary)		Childcare (primary and secondary)		Time with Children		Paid Work		Domestic Labour	
	Father	Mother	Father	Mother	Father	Mother	Father	Mother	Father	Mother
Constant term	0.86***	2.64***	1.40***	5.86***	6.83***	11.84***	5.10***	0.90**	1.08***	3.56***
	(0.14)	(0.24)	(0.29)	(0.48)	(0.40)	(0.46)	(0.41)	(0.30)	(0.19)	(0.23)
Aged										
25-34	-0.08	0.18	-0.24	0.55*	-0.34	0.31	0.66**	-0.35	-0.02	-0.33**
	(0.08)	(0.12)	(0.18)	(0.24)	(0.25)	(0.23)	(0.28)	(0.18)	(0.12)	(0.12)
45-54	-0.12	-0.08	-0.31	0.09	0.41	-0.80	-1.17**	-0.13	0.15	0.02
	(0.10)	(0.23)	(0.22)	(0.45)	(0.30)	(0.43)	(0.3)	(0.34)	(0.15)	(0.23)
Qualifications										
Postgraduate	0.37**	0.61**	1.22***	1.19**	1.02**	0.46	-0.28	1.16**	-0.18	-0.55*
	(0.15)	(0.24)	(0.31)	(0.46)	(0.42)	(0.45)	(0.48)	(0.38)	(0.21)	(0.25)
Bachelor degree	0.37**	0.28	1.24***	1.10**	1.07**	0.12	0.37	0.82**	-0.02	-0.57**
	(0.12)	(0.19)	(0.25)	(0.37)	(0.34)	(0.35)	(0.38)	(0.29)	(0.16)	(0.19)
University diploma	0.16	0.66***	0.62**	1.49***	0.51	0.13	0.17	-0.24	0.06	-0.28
	(0.12)	(0.17)	(0.24)	(0.33)	(0.34)	(0.32)	(0.38)	(0.26)	(0.16)	(0.17)
Skilled vocational	0.10	0.41**	0.28	0.92**	0.00	0.20	0.88**	0.38	0.06	-0.36*
	(0.08)	(0.15)	(0.18)	(0.30)	(0.24)	(0.29)	(0.27)	(0.23)	(0.12)	(0.16)
Basic vocational	0.02	0.02	0.90	0.25	0.29	0.59	0.66	0.54	-0.13	0.10
	(0.24)	(0.23)	(0.52)	(0.44)	(0.71)	(0.43)	(0.78)	(0.32)	(0.33)	(0.21)
Usual hours worked by spouse	0.00*	0.01	0.02***	0.01	0.02**	0.03***	0.00	-0.01*	0.00	0.00
	(0.00)	(0.00)	(0.00)	(0.01)	(0.01)	(0.01)	(0.01)	(0.00)	(0.00)	(0.00)
Labour force status										
Full time		-0.27		-1.15***		-0.81***				
		(0.15)		(0.29)		(0.28)				
Part time	0.54**		1.30***		1.55**				0.07	-0.85***
	(0.16)		(0.34)		(0.47)				(0.23)	(0.15)
Not employed	0.58***	0.66***	1.51***	0.72**	3.34***	1.05***			0.92***	0.40**
	(0.13)	(0.12)	(0.28)	(0.25)	(0.38)	(0.24)			(0.18)	(0.13)
Day of the week										
Saturday	0.23*	-0.27	1.07***	0.40	3.91***	2.11***	-4.78***	-1.74***	1.09***	0.23
	(0.10)	(0.15)	(0.21)	(0.30)	(0.29)	(0.29)	(0.33)	(0.23)	(0.14)	(0.15)
Sunday	0.47***	-0.41**	1.62	0.18	5.10***	2.16***	-5.81	-1.76***	1.32***	0.47**
	(0.10)	(0.14)	(0.20)	(0.28)	(0.28)	(0.27)	(0.31)	(0.22)	(0.13)	(0.15)
Total weekly income	0.00	0.00	0.00**	0.00	0.00	0.00	0.00***	0.00***	0.00	0.00
	(0.00)	(0.00)	(0.00)	(0.00)	(0.00)	(0.00)	(0.00)	(0.00)	(0.00)	(0.00)
Hours child usually attends day care	0.00	-0.02***	-0.02	-0.06***	-0.04***	-0.08***	0.01	0.06***	0.00	-0.01**
	(0.00)	(0.01)	(0.01)	(0.01)	(0.01)	(0.01)	(0.01)	(0.01)	(0.00)	(0.00)
Disabled person in household	0.02	-0.01	0.17**	-0.36	0.57**	0.04	-0.26	-0.50**	0.32**	0.10
	(0.08)	(0.12)	(0.17)	(0.23)	(0.23)	(0.22)	(0.26)	(0.18)	(0.11)	(0.12)
Number of children										
Two	0.19*	0.40**	0.45	0.58**	-0.42	-0.12	0.85**	-0.05	-0.35**	0.21
	(0.08)	(0.13)	(0.18)	(0.25)	(0.24)	(0.24)	(0.27)	(0.19)	(0.12)	(0.13)
Three or more	0.01	0.38**	0.14***	0.35	-0.19	0.30	0.55	-0.52*	-0.12	0.60***
	(0.10)	(0.14)	(0.20)	(0.28)	(0.28)	(0.27)	(0.31)	(0.22)	(0.13)	(0.14)
Youngest child is aged 5-11	-0.70***	-1.93***	-1.26***	-3.55***	-1.29***	-2.86***	0.93**	0.94***	0.01	0.02
	(0.09)	(0.13)	(0.18)	(0.26)	(0.25)	(0.25)	(0.28)	(0.19)	(0.12)	(0.13)
R square	.104	.263	.139	.243	.311	.251	.291	.194	.120	.117

* P-value<0.05. ** P-value<0.01. *** P-value<0.001. N=2926. Std error in brackets

Source: ABS TUS 1997.

Table A14 **Coefficients of minutes a day spent by mothers and fathers in couple-headed households in child care sub-categories (primary activity, shorter model)**

	Physical care		Interactive care		Travel/communication		Passive care	
	Father	Mother	Father	Mother	Father	Mother	Father	Mother
Constant term	10.85**	73.18***	22.59***	50.45***	1.92	10.17*	16.84***	25.18***
	(4.04)	(11.10)	(4.12)	(5.63)	(2.50)	(5.07)	(4.05)	(6.17)
Age								
25-34	2.43	25.56***	0.99	-2.42	-0.93	-3.30	-7.95**	-10.71**
	(2.51)	(5.61)	(2.56)	(2.84)	(1.55)	(2.56)	(2.51)	(3.12)
45-54	-1.15	-8.36	-4.01	-5.69	0.77	-0.25	-2.81	8.83
	(3.08)	(10.31)	(3.14)	(5.23)	(1.90)	(4.71)	(3.09)	(5.73)
Qualification								
Postgraduate	2.27	28.11**	9.41*	12.05*	3.11	-2.56	5.75	-2.26
	(4.32)	(10.73)	(4.41)	(5.44)	(2.67)	(4.90)	(4.33)	(5.96)
Bachelor degree	11.79**	11.76	9.71**	5.09*	2.14	2.78	-0.06	-4.92
	(3.46)	(8.45)	(3.53)	(4.28)	(2.14)	(3.86)	(3.47)	(4.69)
University diploma	4.24	20.06**	7.04*	10.60**	-0.96	2.45	0.03	8.10
	(3.43)	(7.65)	(3.49)	(3.88)	(2.12)	(3.49)	(3.44)	(4.25)
Skilled vocational	0.86	28.42***	2.43	1.43	1.24	-6.59*	2.08	-0.33
	(2.48)	(7.03)	(2.53)	(3.56)	(1.53)	(3.21)	(2.48)	(3.91)
Basic vocational	9.28	17.46	-1.84	6.20	-5.25	-14.07**	-1.19	-8.23
	(7.24)	(10.27)	(7.38)	(5.20)	(4.48)	(4.69)	(7.26)	(5.70)
Usual hours worked by spouse	0.14*	0.26	0.00	-0.15	0.09*	0.29***	0.04	-0.04
	(0.07)	(0.18)	(0.07)	(0.09)	(0.04)	(0.08)	(0.07)	(0.10)
Labour force status								
Full time		6.45		-5.99		-11.96***		-6.70
		(6.78)		(3.44)		(3.10)		(3.77)
Part time	12.34**		7.35		1.83		10.61*	
	(4.83)		(4.92)		(2.98)		(4.84)	
Not employed	8.58*	29.60***	18.09***	3.99	4.25	2.56	3.03	0.39
	(3.89)	(5.67)	(3.96)	(2.87)	(2.40)	(2.59)	(3.90)	(3.15)
Day of the week								
Saturday	4.54	-6.39	4.90	2.35	-1.11	-18.04***	5.27	7.25
	(3.00)	(6.94)	(3.06)	(3.52)	(1.86)	(3.17)	(3.01)	(3.86)
Sunday	6.64**	-5.89	14.16***	-0.53	-1.10	-24.28***	7.13**	6.15
	(2.83)	(6.51)	(2.89)	(3.30)	(1.75)	(2.97)	(2.84)	(3.62)
Total weekly income	0.00	0.00	0.00	0.00	0.00	0.00	0.00	0.00
	(0.00)	(0.01)	(0.00)	(0.00)	(0.00)	(0.00)	(0.00)	(0.00)
Hours child usually attends day care	-0.02	-1.19***	-0.09	-0.18	0.11	0.19	-0.23*	-0.08
	(0.10)	(0.23)	(0.10)	(0.12)	(0.06)	(0.10)	(0.10)	(0.13)
Disabled person in household	3.43	-1.86	-3.28	-4.72	2.23	0.13	-1.73	6.21*
	(2.33)	(5.40)	(2.37)	(2.74)	(1.44)	(2.46)	(2.33)	(3.00)
Number of children								
Two	5.92**	12.10*	0.84	-1.17	2.70	11.70***	1.28	3.10
	(2.48)	(5.76)	(2.53)	(2.92)	(1.54)	(2.63)	(2.49)	(3.20)
Three or more	4.14	2.36	-6.47*	-5.25	4.97**	20.75***	-2.11	6.96
	(2.83)	(6.52)	(2.89)	(3.31)	(1.75)	(2.98)	(2.84)	(3.62)
Youngest child is aged 5-11	-19.30***	-83.12***	-13.92***	-17.54***	1.04	3.45	-10.58***	-18.95***
	(2.58)	(5.96)	(2.62)	(3.02)	(1.59)	(2.72)	(2.58)	(3.31)
R square	.087	.261	.078	.058	.023	.133	.031	.049

* P-value<0.05 ** . P-value<0.01. *** P-value<0.001. N=2926. Std error in brackets
Source: ABS TUS 1997.

Table A15 **Coefficients of minutes a day spent by mothers and fathers in couple-headed households in child care sub-categories (primary and secondary activity, shorter model)**

	Physical Care		Interactive Care		Travel/ Communication		Passive Care	
	Father	Mother	Father	Mother	Father	Mother	Father	Mother
Constant term	12.37**	75.61***	34.98***	88.85***	2.23	12.03*	44.02**	202.02***
Aged	(4.39)	(11.77)	(7.14)	(13.51)	(2.57)	(5.32)	(15.48)	(28.76)
25-34	2.93	26.24***	0.93	-11.31	-1.10	-4.08	-20.16*	32.23*
	(2.72)	(5.95)	(4.43)	(6.83)	(1.60)	(2.69)	(9.61)	(14.53)
45-54	-1.67	-7.46	-5.85	-22.21	1.13	-0.72	-14.93	22.57
	(3.34)	(10.93)	(5.44)	(12.55)	(1.96)	(4.95)	(11.80)	(26.71)
Qualification								
Postgraduate	4.67	30.19**	20.55**	26.23*	2.80	-3.32	50.63**	34.22
	(4.69)	(11.37)	(7.64)	(13.05)	(2.75)	(5.14)	(16.57)	(27.79)
Bachelor degree	13.13***	13.67*	18.36**	13.04*	2.30	2.45	44.42	50.53*
	(3.76)	(8.95)	(6.12)	(10.28)	(2.20)	(4.05)	(13.27)	(21.89)
University diploma	5.08	28.18**	9.73	2.98	-0.72	2.37	25.29	65.98**
	(3.72)	(8.11)	(6.06)	(9.31)	(2.18)	(3.67)	(13.14)	(19.82)
Skilled vocational	1.44	29.04***	3.09	2.81	1.07	-5.27	12.13	39.19*
	(2.69)	(7.45)	(4.38)	(8.56)	(1.58)	(3.37)	(9.50)	(18.21)
Basic vocational	13.10	20.50	9.07	-9.51	-5.36	-14.98**	35.02	32.57
	(7.86)	(10.88)	(12.80)	(12.49)	(4.61)	(4.92)	(27.75)	(26.60)
Usual hours worked by spouse	0.14	0.39*	0.06	0.18	0.12**	0.27**	0.83**	0.27
	(0.08)	(0.19)	(0.12)	(0.21)	(0.04)	(0.08)	(0.27)	(0.45)
Labour Force Status								
Full time		6.01		-25.14**		-10.43**		-48.58**
		(7.19)		(8.25)		(3.25)		(17.56)
Part time	15.39		11.48		1.05		53.02**	
	(5.24)		(8.53)		(3.07)		(18.49)	
Not employed	9.73*	30.39***	19.89**	4.55	4.22	2.52	62.58***	11.52
	(4.22)	(6.01)	(6.87)	(6.90)	(2.47)	(2.72)	(14.90)	(14.69)
Total weekly income	0.01	0.00	0.00	0.02**	0.00	0.00	0.02**	0.01
	(0.00)	(0.01)	(0.01)	(0.01)	(0.00)	(0.00)	(0.01)	(0.02)
Day of the week								
Saturday	5.12	-7.24	16.40**	-6.06	-0.20	-16.78***	47.13***	58.34**
	(3.26)	(7.36)	(5.30)	(8.45)	(1.91)	(3.33)	(11.50)	(17.99)
Sunday	7.40**	-8.17	29.25***	-13.47	-1.26	-24.81***	64.81***	62.94***
	(3.07)	(6.90)	(5.00)	(7.92)	(1.80)	(3.12)	(10.85)	(16.86)
Hours child usually attends day care	-0.03	-1.28***	-0.20	-1.07***	0.09	0.15	-0.93	-2.23***
	(0.11)	(0.24)	(0.18)	(0.28)	(0.06)	(0.11)	(0.38)	(0.59)
Disabled person in household	3.72	-0.93	-0.14	-8.50	2.33	-0.29	3.15	-15.09
	(2.53)	(5.72)	(4.11)	(6.57)	(1.48)	(2.59)	(8.92)	(13.98)
Number of children								
Two	5.10	13.62*	5.61	20.24**	3.19*	12.85***	15.46	4.99
	(2.70)	(6.11)	(4.39)	(7.01)	(1.58)	(2.76)	(9.51)	(14.93)
Three or more	3.16	7.91	-1.79	23.00**	5.78**	22.26***	1.52	-16.43
	(3.08)	(6.91)	(5.01)	(7.93)	(1.80)	(3.13)	(10.86)	(16.89)
Youngest child is aged 5-11	-21.43***	-89.48***	-15.67**	-30.15***	0.68	2.65	-49.48	-136.83***
	(2.80)	(6.32)	(4.55)	(7.26)	(1.64)	(2.86)	(9.87)	(15.45)
R square	.092	.263	.057	.064	.026	.123	.082	.129

* P-value<0.05 **. P-value<0.01. *** P-value<0.001. N=2926. Std error in brackets
Source: ABS TUS 1997.

Table A16   Coefficients of hours a day spent by mothers and fathers in couple-headed households

	Childcare (primary and secondary) alone with children		Time alone with children		Childfree leisure		Leisure excluding simultaneous childcare time	
	Father	Mother	Father	Mother	Father	Mother	Father	Mother
Constant term	0.35***	1.90***	0.74***	3.21***	0.61	0.01	2.95**	1.64***
	(0.13)	(0.33)	(0.18)	(0.46)	(0.15)	(0.11)	(0.37)	(0.25)
Age								
25-34	-0.21**	0.54**	-0.23*	0.79**	-0.11	-0.03	-0.18	0.07
	(0.08)	(0.17)	(0.11)	(0.23)	(0.09)	(0.06)	(0.07)	(0.01)
45-54	0.07	-0.49	0.04	-1.06*	0.18	-0.04	0.52**	-0.32
	(0.10)	(0.31)	(0.14)	(0.42)	(0.11)	(0.10)	(0.30)	(0.14)
Qualifications								
Postgraduate	0.40**	0.29	0.64**	-0.19	-0.44**	-0.14	-0.74**	-0.35
	(0.14)	(0.32)	(0.19)	(0.44)	(0.16)	(0.11)	(0.37)	(0.39)
Bachelor degree	0.16	0.58*	0.12	0.49	-0.44*	0.03	-0.97***	-0.21
	(0.11)	(0.25)	(0.16)	(0.35)	(0.13)	(0.09)	(0.13)	(0.12)
University diploma	-0.03	0.19	-0.14	-0.16	-0.18	0.06	-0.32	-0.49**
	(0.11)	(0.23)	(0.15)	(0.31)	(0.13)	(0.08)	(0.02)	(0.10)
Skilled vocational	0.06	0.03	0.02	-0.60*	-0.28	0.11	-0.53**	0.01
	(0.08)	(0.21)	(0.11)	(0.29)	(0.09)	(0.07)	(0.21)	(0.06)
Basic vocational	0.03	0.14	0.01	0.54	-0.44	-0.03	-0.72	0.24
	(0.24)	(0.31)	(0.32)	(0.42)	(0.27)	(0.10)	(0.32)	(0.27)
Usual hours worked by spouse	0.01***	0.03***	0.02***	0.06***	0.00	0.00	0.00	0.00
	(0.00)	(0.01)	(0.00)	(0.01)	(0.00)	(0.00)	(0.00)	(0.00)
Labour force status								
Full time		-0.99***		-1.28***		-0.21**		-0.15
		(0.20)		(0.28)		(0.07)		(0.05)
Part time	0.54**		0.78***		0.04		-0.10	
	(0.16)		(0.22)		(0.18)		(0.03)	
Not employed	0.42**	0.34*	0.65***	0.45	0.47*	0.15**	1.77***	0.43*
	(0.13)	(0.17)	(0.17)	(0.23)	(0.14)	(0.06)	(0.17)	(0.14)
Day of the week								
Saturday	-0.02	-1.17***	0.17	-1.80***	0.32*	0.05**	1.30***	0.00
	(0.10)	(0.21)	(0.13)	(0.29)	(0.11)	(0.07)	(0.12)	(0.41)
Sunday	0.11	-1.50***	0.28*	-2.50***	0.03	-0.17	1.19***	0.66***
	(0.09)	(0.19)	(0.13)	(0.27)	(0.10)	(0.07)	(0.21)	(0.05)
Total weekly income	0.00	0.00*	0.00*	0.00	0.00	0.00	0.00	0.70***
	(0.00)	(0.00)	(0.00)	(0.00)	(0.00)	(0.00)	(0.00)	(0.35)
Hours child usually attends day care	-0.01*	-0.03***	-0.01*	-0.04***	0.00	0.00	-0.01	0.01
	(0.00)	(0.01)	(0.00)	(0.01)	(0.00)	(0.00)	(0.00)	(0.01)
Disabled person in household	0.08	-0.31*	0.12	-0.39	-0.04	0.08	0.02	-0.11
	(0.08)	(0.16)	(0.10)	(0.22)	(0.09)	(0.05)	(0.05)	(0.10)
Number of children								
Two	0.02	0.12	0.14	0.16	-0.04	-0.02	-0.48**	-0.29*
	(0.08)	(0.17)	(0.11)	(0.24)	(0.09)	(0.06)	(0.05)	
Three or more	-0.08	-0.24	0.02	-0.33	-0.03	-0.05	-0.18	0.04
	(0.09)	(0.19)	(0.13)	(0.27)	(0.10)	(0.07)	(0.14)	(0.11)
Youngest child 5-11	-0.50***	-1.89***	-0.57***	-2.46***	0.01**	0.38	0.11	0.81***
	(0.08)	(0.18)	(0.12)	(0.24)	(0.10)	(0.06)	(0.06)	(0.12)
R square	.064	.275	.017	.277	.030	.026	.043	.021

* P-value<0.05. ** P-value<0.01. *** P-value<0.001. N=2926. Std error in brackets
Source: ABS TUS 1997.

Table A17   Cross-national regression models

Predictor Variables	Model 1		Model 2		Model 3		Model 4	
	Total Work	Unpaid Work	Total Work	Unpaid Work	Total Work	Unpaid Work	Total Work	Unpaid Work
Constant	9.16	1.59	9.70	1.90	9.64	1.78	9.63	2.12
Country Effects								
Norway	-0.36 ***	-0.32 ***						
	(0.08)	(0.07)						
Germany	0.16 **	0.12 *						
	(0.06)	(0.05)						
Italy	-0.33 ***	-0.42 ***						
	(0.06)	(0.05)						
Sex								
Female	0.65 ***	3.08 ***						
	(0.04)	(0.04)						
Sex and country								
Female								
Australian			-0.36 ***	2.51 ***	-0.34 ***	2.56 ***	-0.48 ***	1.66 ***
			(0.08)	(0.07)	(0.08)	(0.07)	(0.12)	(0.10)
German			-0.22 **	2.59 ***	-0.32 ***	2.46 ***	-0.37 **	1.80 ***
			(0.08)	(0.07)	(0.08)	(0.07)	(0.11)	(0.10)
Norwegian			-0.71 ***	1.76 ***	-0.76 ***	1.70 ***	-0.41 *	1.08 ***
			(0.11)	(0.09)	(0.11)	(0.09)	(0.16)	(0.10)
Italian			0.46 ***	2.84 ***	0.47 ***	2.87 ***	0.52 ***	2.65 ***
			(0.08)	(0.07)	(0.08)	(0.07)	(0.12)	(0.10)
Male								
German			0.23 **	0.13	0.13	-0.04	0.12	-0.04
			(0.08)	(0.07)	(0.08)	(0.07)	(0.12)	(0.14)
Norwegian			-0.26 *	0.19 *	-0.29 **	0.14	-0.46 *	0.03
			(0.11)	(0.09)	(0.11)	(0.09)	(0.17)	(0.10)
Italian			-1.67 ***	-1.23 ***	-1.67 ***	-1.24 ***	-1.52 ***	-1.08 ***
			(0.08)	(0.07)	(0.08)	(0.07)	(0.12)	(0.10)
Age of youngest child								
< 5					0.73 ***	1.24 ***	0.66 **	0.61 ***
					(0.07)	(0.06)		(0.12)
5-11					0.20 **	0.38 ***	0.49 ***	-0.02
					(0.07)	(0.06)	(0.17)	(0.14)
Female and youngest child under 5								
Australian							1.07 ***	2.67 ***
							(0.14)	(0.12)
Italian							0.71 ***	1.25 ***
							(0.12)	(0.10)
Norwegian							0.28	2.11 ***
							(0.20)	(0.16)
German							0.87 ***	2.18 ***
							(0.11)	(0.10)

Table A17   Continued

Predictor Variables	Model 1		Model 2		Model 3		Model 4	
	Total Work	*Unpaid Work*	*Total Work*	*Unpaid Work*	*Total Work*	*Unpaid Work*	*Total Work*	*Unpaid Work*
Male and youngest child under 5								
Australian							0.66 ***	0.62 ***
							(0.15)	(0.12)
Italian							0.51 ***	0.39 ***
							(0.12)	(0.10)
Norwegian							1.17 ***	0.95 ***
							(0.20)	(0.17)
German							0.79 ***	0.72 ***
							(0.12)	(0.10)
Female and youngest child 5 to 11								
Australian							0.52 **	1.45 ***
							(0.17)	(0.14)
Italian							0.16	0.41 ***
							(0.10)	(0.09)
Norwegian							-0.41	0.64 **
							(0.22)	(0.19)
Male and youngest child 5 to 11								
Australian							0.50 **	-0.02
							(0.17)	
Italian							0.08	-0.11
							(0.11)	(0.09)
Norwegian							0.31	0.04
							(0.26)	(0.21)

Table A17 Continued

Control Variables	Model 1		Model 2		Model 3		Model 4	
	Total Work	*Unpaid Work*	*Total Work*	*Unpaid Work*	*Total Work*	*Unpaid Work*	*Total Work*	*Unpaid Work*
Age								
25 to 34	0.11 *	0.16 ***	0.08	0.14 ***	-0.09 ***	-0.14 ***	-0.10 *	-0.23 ***
	(0.04)	(0.04)	(0.04)	(0.04)	(0.05)	(0.04)	(0.05)	(0.04)
45 to 54	0.02	0.15 **	-0.02	0.13 **	0.10	0.34 ***	0.10	0.33 ***
	(0.05)	(0.04)	(0.05)	(0.04)	(0.05)	(0.05)	(0.06)	(0.05)
Education								
Above secondary	0.11	0.03	0.05	0.00	0.01	-0.07	0.00	-0.09 *
	(0.06)	(0.05)	(0.06)	(0.05)	(0.06)	(0.05)	(0.06)	(0.05)
No of children under 18	0.34 ***	0.51 ***	0.34 ***	0.51 ***	0.19	0.25 ***	0.18 ***	0.23 ***
	(0.02)	(0.02)	(0.02)	(0.02)	(0.03)	(0.02)	(0.03)	(0.02)
Day of week								
Saturday	-2.21 ***	0.73 ***	-2.22 ***	0.72 ***	-2.21 ***	0.73 ***	-2.22 ***	0.73 ***
	(0.05)	(0.04)	(0.05)	(0.04)	(0.05)	(0.04)	(0.05)	(0.04)
Sunday	-4.91 ***	-0.34 ***	-4.91 ***	-0.35 ***	-4.91 ***	-0.34 ***	-4.91 ***	-0.34 ***
	(0.05)	(0.04)	(0.05)	(0.04)	(0.05)	(0.04)	(0.05)	(0.04)
Work Status								
Part time		0.76 ***		1.15 ***		1.12 ***		0.95 ***
		(0.06)		(0.06)		(0.06)		(0.06)
Not employed		2.34 ***		2.20 ***	***	2.17 ***		2.06 ***
		(0.04)		(0.04)		(0.04)		(0.04)
Income								
Lowest quartile	-0.72 ***	-0.20 **	-0.75 ***	-0.18 *	-0.75	-0.16 *	-0.73 ***	-0.14
	(0.09)	(0.07)	(0.09)	(0.07)	(0.09)	(0.07)	(0.09)	(0.07)
Highest quartile	0.14 *	-0.23 ***	0.18 **	-0.23 ***	0.22 ***	-0.16 ***	0.23 ***	-0.13 **
	(0.06)	(0.05)	(0.06)	(0.05)	(0.06)	(0.05)	(0.06)	(0.05)
Spouse's Work								
Part time	0.26 ***	0.30 ***	0.07	0.06	-0.01	-0.08	-0.05	-0.03
	(0.07)	(0.06)	(0.07)	(0.06)	(0.07)	(0.06)	(0.07)	(0.06)
Not in labour force	-0.71 ***	-0.45 ***	-0.47 ***	-0.28 ***	-0.52 ***	-0.37 ***	-0.54 ***	-0.31 ***
	(0.05)	(0.04)	(0.05)	(0.04)	(0.05)	(0.04)	(0.05)	(0.04)
R square	.342	.446	.370	.478	.373	.489	.374	.364

* P-value<0.05. ** P-value<0.01. *** P-value<0.001. N=29767. Std error in brackets.
Source: MTUS World 5.1.

Index